To Denis Vaughan
from Alan Jordan
October 1980

ACOUSTICAL DESIGN OF
CONCERT HALLS AND THEATRES

Acoustical Design of Concert Halls and Theatres

A Personal Account

VILHELM LASSEN JORDAN, M.Sc., Ph.D.

Acoustic Consultant, Gevninge, DK 4000 Roskilde, Denmark

APPLIED SCIENCE PUBLISHERS LTD
LONDON

APPLIED SCIENCE PUBLISHERS LTD
RIPPLE ROAD, BARKING, ESSEX, ENGLAND

British Library Cataloguing in Publication Data

Jordan, Vilhelm Lassen
    Acoustical design of concert halls and theatres.
    1. Architectural acoustics
    2. Auditoriums
    I. Title
    725'.83   NA2800

    ISBN 0-85334-853-7

WITH 14 TABLES AND 168 ILLUSTRATIONS

Printed in Great Britain by Galliard (Printers) Ltd, Great Yarmouth

Foreword

About fifteen years ago I and my colleagues at the University of Salford were organising a three-month postgraduate course in acoustics. Before the course arrangements were finalised I visited Dr Per V. Bruel of Bruel & Kjaer A/s whilst on holiday in Denmark in order to invite him to contribute. As a result of our discussions he suggested that I really ought to try to persuade Dr Vilhelm L. Jordan to take part, who, I was told, was very knowledgeable on auditorium acoustics and had a great deal to offer to students who were seeking authoritative information in this particular field. In my ignorance, I had no appreciation of Dr Jordan's standing or reputation in acoustics, being a newcomer myself to the subject at that time.

Introductions were duly made for me and I was invited to Dr Jordan's home at Gevninge, near Roskilde, for dinner. After a pleasant social evening I eventually broached the subject of his possible contribution to our course. I explained that if he could help at all it would be in the capacity of a lecturer and as a tutor in the experimental laboratory. His reaction was that he would give the proposal careful consideration and let me have his decision within 24 hours. A telephone call followed and much to my delight and satisfaction the gist of the conversation was that my invitation was accepted and he was prepared to come and give a series of lectures in Salford and would spend a fortnight with us, devoting some of the time to supervision of students in the laboratory. Furthermore, he would only present a bill for his travelling expenses, a gesture which, as the years pass by, I have come to appreciate as typical of his generosity, particularly where young people are involved.

Dr Jordan's efforts on our behalf did not end there because several weeks before the course was due to commence, a large parcel of lecture notes arrived with instructions to turn them into presentable English (not that this required very much effort). The notes were the basis of ten lectures on auditorium acoustics; not exclusively dealing with acoustics, but also describing the historical development of auditoria.

My memory of those lectures and the wide-ranging and rather unique personal experience which Dr Jordan has had in acting as an acoustic consultant in many parts of the world made me decide to try to persuade him to write a book which summarised that experience. After a little pressure he capitulated and this book is the result.

I accept that there are one or two good textbooks available on auditorium acoustics which contain all the necessary basic mathematics and science for a thorough grounding in the subject, but there are very few which present a personal view based on close involvement with a large number of concert hall and theatre projects. I feel that the book which has been produced has a lot to offer to the architect, theatre consultant and acoustician who want to know how this or that problem was approached and whether the final solution was a success or not. Students who are studying acoustics as part of a degree course or involved in postgraduate study will also find much that will interest them.

PETER LORD
Salford University

Preface

When the publishers first approached me about this volume I hesitated to accept. To put it frankly, I did not consider myself particularly well qualified to write a textbook on Acoustics. Previously, I had written a review for the journal *Applied Acoustics* and I offered to bring this review up to date. After I had done that and the publishers still insisted on the book I gave in, but I explained that it might end up as something quite different from a textbook. To be brief, having accepted the commission I am grateful to the publishers for the opportunity to present a personal account.

I am also grateful to their professional adviser on Acoustics, my good friend Professor Peter Lord, who, I guess, was behind this enterprise from the very beginning. We have had a long acquaintanceship during which I also have had the privilege, of lecturing in his department at the University of Salford; in fact, those lectures could be regarded as the first step towards this venture.

Looking back I must mention the names of the two men who introduced me to the subject of Acoustics: the late P. O. Pedersen, Professor in Telecommunications and for many years the Director of the Royal Technical University of Copenhagen, and the late Erwin Meyer, Professor of Acoustics and the departmental head of one of the first institutes that adopted technical acoustics as an independent branch of research, the Heinrich Hertz Institute of Berlin. After the war, Erwin Meyer was for many years the inspiring leader of the 3rd Institute of Physics of the University of Göttingen. Many years ago P. O. Pedersen proposed to study acoustic problems using models (the basis for my doctoral thesis) and Erwin Meyer introduced me to the fundamental problems of room acoustics and the associated methods of measurement. Professor Pedersen arranged for me to join the staff of the technical department at the Danish Broadcasting Corporation during the design and construction period of 'Radiohuset', taking charge of acoustical measurements.

It is impossible to quote a comprehensive list of all the professional colleagues with whom I have had profitable discussions on matters pertaining to the subject of this book, but it is appropriate to name two associations of lasting importance, W. Reichardt of Dresden and M. R. Schroeder of Göttingen, neither of whom needs further introduction. Both of them share a common interest with the author in acoustical criteria for concert halls.

Collaboration with architects, consulting engineers and other consultants has in general been very fruitful and so have the joint endeavours to achieve the best results in acoustics and noise reduction in the many projects in which I have been involved. Some of the names of architects and other consultants are mentioned in the text but many other people, including builders, contractors and workers, are anonymous.

My permanent collaborator and business associate, Niels V. Jordan, has conducted many of the measurements quoted and has prepared all the data sheets, illustrations and pictures.

The credit for typing of the manuscript and, along with this task, many suggestions on style and presentation is due to my wife, Ebba Jordan.

VILHELM LASSEN JORDAN

Contents

Foreword v
Preface vii
Introduction xi
Abbreviations and Symbols xiii

1 Early Ventures 1
2 Brief Historical Survey 22
3 Multipurpose Halls—Early Period 39
4 Development of Criteria and Model Research, I . . . 57
5 New York State Theatre and Metropolitan Opera House, and
   National Theatres of Latin America 75
6 Sydney Opera House 92
7 Oslo Concert Hall—Design Development 122
8 Multiform and Multipurpose Halls 139
9 Development of Criteria and Model Research, II . . 158
10 Recent Projects of Varying Design 165
11 Development of Criteria and Model Research, III . . 187
12 Natural and Artificial Acoustics 197

Appendix I   Building-up Process, Rise Time and Inversion Index   206
Appendix II   Building-up Process, Steepness and Inversion Index   208
Appendix III Decay Process, EDT and Inversion Index . .   212
Appendix IV   Impulse Response Level as Acoustical Criterion .   216
Appendix V   Expectation Values of C, RR and LE, and Measured
              Values of LE 217

Index 221

Introduction

The acoustic problems associated with the design of large halls are by no means a new subject of interest. Throughout the 19th century the problems attracted growing attention, and it is obvious that they grew with increasing size of the auditoria. Nevertheless, the interest in this subject has increased even more in our own century, and the reasons are evident: the size of the auditoria has continued to increase, the architecture has developed with new shapes and materials and communication technology has provided powerful tools (physical, physiological and psychological) for investigating acoustics and has also provided many possibilities of creating artificial acoustical environments.

The acoustics of large halls of the 19th and earlier centuries were the result of tradition and very varied, but the new trends of the 20th century have been on the one hand to use a scientific approach and on the other an artistic, architectural approach. It cannot be denied that the practical results of these endeavours bear witness to an insufficient reconciliation between these two trends, and that we have reached a critical point in the development of the art and science of applied room acoustics. However, it is also a point at which we are beginning to gain a better understanding of essential design features and of the inherent acoustic advantages of the 19th century halls.

The reader should not expect a systematic treatment of the subject, but rather an approach which is intimately linked with the author's own experience, as an acoustic consultant for over 30 years, an experience closely associated with a number of individual projects he has been occupied with (as projects and as completed halls). It is also an experience which is linked with his own studies of acoustical criteria and their development.

An especially powerful tool, assisting in these studies, has been the work with scaled acoustic models, which have been applied, intensively, over a long period. Mostly, the studies have been confined to objective testing of physical parameters. This course has been chosen partly because of the intricacy and the ensuing impracticality of extending the frequency range to cover the whole audible range, and also partly because the methods of subjective assessment of acoustic qualities (of musical samples) are suitable for basic research but not for the evaluation of individual projects.

It may seem a bit odd to include in this volume such early ventures as a broadcasting house of 1945 and multipurpose halls of the 1950s. Nevertheless, it may be possible that even today there will be benefits to gain, both from the errors and from the positive experience of that distant period. As the manuscript developed it even became necessary to include a brief historical survey of the design development for theatre design as well as concert hall design.

The idea of editing a series of 'case histories', which eventually turned out to be the only possible way open to the author to express himself properly, had to be supplemented and interwoven with chapters presenting the development of acoustical criteria and of model research. These chapters also attempt to take into account contemporary contributions from several other authors (without being complete in any way) and they may be considered as an updated version (1976–78) of the author's earlier publications on the same subject in *Applied Acoustics*.

The main theme throughout the volume has been the 'natural' acoustics of auditoria and only limited comments on 'artificial' acoustics are added. The various loudspeaker systems of the individual halls are briefly mentioned in connection with those halls.

Theoretical considerations and development of some formulae are given in appendices.

References to papers and publications are given at the end of each chapter without indicating the specific association with the text.

BIBLIOGRAPHY

Subject	Reference
Development of acoustical criteria and of model research	Jordan, V. L. *Applied Acoustics*, **2**, 1969, p. 77.
	Jordan, V. L. *Applied Acoustics*, **8**, 1975, p. 217.

Abbreviations and Symbols

RT (s)   Reverberation Time—time interval corresponding to 60 dB level reduction in a reverberation process.

TR (ms)   Rise Time—time interval corresponding to arrival of 50% of the energy in a build-up process, or, expressed in dB: -3 dB below final level.

II (coefficient)   Inversion Index $\sim$ TR (audience area)/TR (stage area). II may also be applied in association with Steepness values or EDT values.

EDT (s)   Early Decay Time—time interval corresponding to 60 dB level reduction, calculated from the slope of the first 10 dB of a reverberation process. Measurements are possible with the integrated pulse method.

σ (dB/ms)   Steepness of a build-up process at the -5 dB point.

σ_{calc} (dB/ms)   Calculated value of Steepness, from value of RT.

c (m/s)   Velocity of sound in atmospheric air.

α (coefficient)   Sound absorption coefficient.

p (m/s)   Mean free path, between successive reflections in a room.

N (dB/octave)   Prefix for noise rating curve according to ISO standards.

C (dB)   Clarity—10 log (energy over first 80 ms expressed as a fraction of the remaining energy).

RR (dB)   Room Response—10 log (lateral energy between 25 and 80 ms plus total energy between 80 and 160 ms expressed as a fraction of the total energy between 0 and 80 ms).

LE (coefficient)   Lateral Efficiency—lateral energy between 25 and 80 ms as a fraction of the total energy between 0 and 80 ms. Lateral energy is measured with a figure-of-eight microphone with maximum sensitivity at right angles to the longitudinal axis of a hall.

C, RR, LE   Expected values are calculated from RT (or EDT) values.

Chapter 1

Early Ventures

Radiohuset, Copenhagen

The Broadcasting House of Copenhagen (Radiohuset) was built during the war and the German occupation. It was a major attempt to provide adequate and up-to-date space and facilities for the State Broadcasting System, with studios for speech and music having good acoustical conditions. At that time, the Danish Broadcasting Corporation had its domicile in a building adjacent to the old Royal Theatre. The building, nicknamed 'Staerekassen'—starling's (nest) box—by a Copenhagen wit, due to its appearance, also contained a 'new stage' for the theatre. It was a building in reinforced concrete where sound isolation requirements had been difficult to fulfil.

For the new Broadcasting House the sound isolation requirements were well defined and were also carefully considered, whereas exact design of room acoustics as applied to studios was generally less well known, due to limited experience. Certain standards for Reverberation Time (RT) for speech and music studios did exist at the time, but the question of frequency dependence of RT and also the proper amount of sound diffusion were largely unsolved problems.

The whole design process, therefore, went along simultaneously with the actual construction, to a certain extent, and especially in the case of the interior fitting of panelling. This happened to be to the obvious benefit of the author as a young acoustician, mostly working on his own and making measurements (of RT v. frequency) of the individual studios at different stages.

The thoughts of the architect (Vilhelm Lauritzen) on the interior of the studios had concentrated on appearance and solidity and he had

1

recommended the application of rather thick (16–22 mm) wooden panels as a standard cover for ceiling and wall surfaces. Sound absorption then became a problem at medium and medium high frequencies (200–2000 Hz), but fortunately the resonance absorption of perforated panelling was recognised, had been investigated and some results had been published.

Resonance absorption provided a suitable, although sometimes complicated, solution to the problem of equalising the sound absorption v. frequency curve. (Some of the studios were designed with no less than 6–10 differently tuned panel systems; the panelling had either circular holes or narrow slots.) In a certain studio of medium size (Studio 3), intended for musical ensembles and for choir, the idea of creating variable RT conditions was tried out. It was not the overall value of RT as much as its frequency dependence which could be varied simply by moving the perforated panels, hinged along one vertical edge. The panels could be turned, within a wooden frame, thus varying the volume behind the panels, thereby also varying their resonance frequency (the frequency of maximum absorption) within corresponding limits (300–1700 Hz).

By arranging the panel areas of the walls in groups which could have different angled positions, a large variety of RT v. frequency characteristics could be easily obtained, especially since the movement of the panel groups was controlled by a pneumatic system, located in the sound control room. The perforated type of panelling was supplemented with a louvre type, providing low frequency absorption when closed and high frequency absorption when opened.

This arrangement was very convenient for testing subjective assessment of different types of RT v. frequency characteristics of a music studio. The testing was arranged by connecting the studio microphones to a monitor speaker in another control room, with no visual contact with the studio. A group of conductors and musically trained people were asked to assess the musical (acoustical) quality of a fixed and repeated musical programme. By

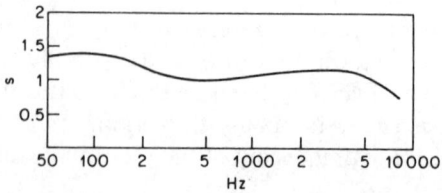

FIG. 1.1.   RT v. frequency curve—optimum selected at musical tests in Studio 3, Radiohuset, Copenhagen, vol. 1400 m³.

repeating the same RT conditions arbitrarily, the capacity of the judges for this kind of assessment could also be tested.

From results of the individual and independently edited written statements, of the many different RT conditions, one in particular was unanimously accepted. These conditions are shown in Fig. 1.1. It is interesting to compare this result with the investigations of Békésy (Fig. 1.2), conducted some years earlier at the premises of the Budapest

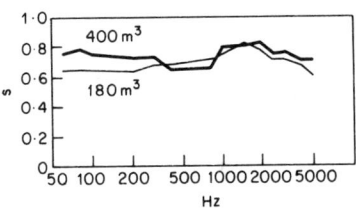

FIG. 1.2.   RT v. frequency curve—optimum found by the tests of Békésy at a Budapest broadcasting studio, vol. 400 m³.

Broadcasting Studios. The 400 m³ studio of Békésy had a preferred RT v. frequency curve which is very close to that for the 1400 m³ studio of Radiohuset, although differing in absolute values.

It is also interesting to note that other studios of Radiohuset (one of 450 m³ and one of 2400 m³) were adjusted several times during the fixing of wall panelling and at the same time subjected to musical testing. Eventually

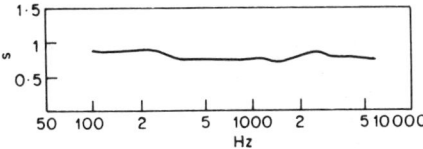

FIG. 1.3.   RT v. frequency curve—optimum, as adjusted during construction of Studio 4, Radiohuset, Copenhagen, vol. 450 m³.

RT v. frequency curves very close to the optimum curve of Studio 3 were achieved, although differing in absolute values (see Figs. 1.3 and 1.4).

These experiences have been a guide on later occasions concerning the essential features of the optimum RT v. frequency dependence, not only of studios but also of concert halls. They could be summarised in that it is especially important to have a 'dip' in the medium frequency region (300–800 Hz) and also that the increase of RT at the higher frequencies is

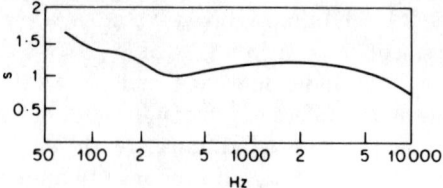

FIG. 1.4.   RT v. frequency curve—optimum, as adjusted during construction of Studio 2, Radiohuset, Copenhagen, vol. 2400 m³.

perhaps more important than the increase at the lower frequencies—contrary to many, recent and often repeated, statements about the importance of the increase at low frequencies.

An objective explanation of these findings would have to consider the energy spectrum of musical instruments and more specifically, in large halls, the energy spectrum of a full orchestra. It is well known that the heavy instruments (brass and percussion) are responsible for the maximum of the energy v. frequency curve (at 100–300 Hz), and it is also well known that one of the most frequently experienced defects occurring in orchestral music is a dominance of the heavy groups over the more subtle contributions from the wood wind and strings, especially in *tuttis*. This dominance becomes

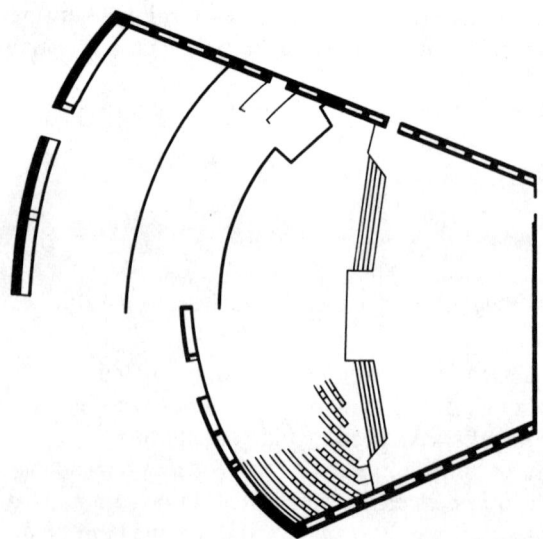

FIG. 1.5.   Plan of the Concert Studio (Studio 1, Radiohuset).

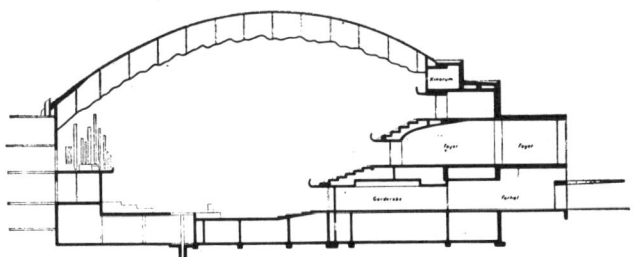

FIG. 1.6. Longitudinal section of the Concert Studio.

especially problematic when full orchestras are playing in undersized halls or studios.

It is not unreasonable to assume that the RT v. frequency dependence should be the inverse of the energy dependence since the stationary energy in a room has a reciprocal dependence on RT.

So far our considerations have only regarded the stationary energy and the statistical process of reverberation. In a later discussion (Chapter 11) we shall follow this up with regard to the influence of early energy, transients and single reflections on the frequency spectrum.

The main studio of Radiohuset is the Concert Studio (Studio 1). It has a seating capacity for about 1000 and it also functions as a concert hall. Both the early history and the later development of this auditorium have features of unusual interest. The main shape (Figs. 1.5, 1.6 and 1.7) had been proposed by the architect and was influenced by acoustical concepts much favoured by architects of that period.

FIG. 1.7. View towards the stage of the Concert Studio.

These concepts could be traced back to the ancient Greek and Roman theatres but they also included some more recent features which were expressed, probably for the first time, by the design of the Salle Pleyel Concert Hall in Paris (constructed in 1927). Each performer was imagined as a point source located in the stage area and the sound was conceived as rays, radiating from the performer and reflected from the boundaries, especially the ceiling and side walls. If the rays should propagate as parallel bundles after being reflected from the ceiling then the ceiling shape should be parabolic (in one or more sections) with a focus area at the position of each performer. The side walls should be angled to distribute the side reflections towards the audience. This tendency to apply and to supplement the ancient *theatre* concepts, originally intended for the *spoken* word, to the design of *concert halls*, mainly used for *symphonic music*, has had a long-lasting and detrimental influence. When considering the design development of concert halls over the last half century it may be true that this influence is now receding but it has not yet disappeared altogether.

Although the shape of the ceiling of the Concert Studio has some resemblance to the ceiling shape of Salle Pleyel the modifications were made for structural and not for acoustical reasons. The shape was defined by the static requirements of a (thin) self supporting reinforced concrete shell (a whim of the period). It might have been envisaged that this somewhat vault-like shape could create focusing problems and probably also modal singularities between ceiling and floor at low frequencies. But the exterior shape of the concrete shell did not necessarily define the interior ceiling shape, since a suspended ceiling had to be provided, for sound isolation and air distribution. The shape of this interior ceiling would be part of the acoustical design of the hall.

The appearance in plan is a very open fan shape with the audience evenly distributed over three levels. The obvious consideration of the design was to distribute efficiently the direct sound and the first reflections from the stage towards the audience.

The large organ is located to the rear of the stage which meant that the width of the stage had to be adapted to the broad organ facade. It also meant that the ceiling over the platform had to start from a high level (defined by the top level of the organ pipes). The erection of the hall took place during the German occupation of Denmark but the completion was deliberately delayed so that any inauguration during that period was out of the question. This meant that there was plenty of time for experimenting with the acoustical properties of the interior and for settling the problematic design points.

One of the major problems was the phenomenon of unexpectedly strong modes at certain low frequencies, between ceiling and floor. They could be demonstrated, and measured, by a simple arrangement consisting of a loud-speaker on a baffle positioned at floor level (with the baffle horizontally orientated), a figure-of-eight microphone suspended and travelling along a vertical path between floor and ceiling, and a pure tone generator feeding the speaker. By applying a gliding frequency (interval 30–200 Hz) the prominent maxima were recorded on a level recorder. By tuning to one of these maxima and by interrupting the tone, the decay process could also be recorded. This procedure was repeated for the various strong modes (some of them having remarkably long decay processes, especially those below 100 Hz. One mode had a decay corresponding to RT ∼ 15 s).

It was suggested that the most expedient way of diffusing these modes, and reducing their decay process, would be to shape the inner ceiling with a series of 'waves'. This treatment had been applied in some of the medium sized studios to avoid standing waves between floor and ceiling and had been tested by similar methods. It had been found that to diffuse effectively, the span of the 'waves' (zig-zag panels) had to be of the same order as the wavelength of prominent modes.

Experimenting with correspondingly larger shapes in full scale in the Concert Studio was out of the question. Instead a small-scale experiment using a 1/50 scale model (in gypsum plaster) of the studio was suggested. This was an early attempt to apply model techniques to the investigation of acoustic problems in auditoria. It meant working with frequencies 50 times higher than the mode frequencies of the Concert Studio, i.e. frequencies in the range of 5–10 kHz. At the time it was just feasible to provide adequate miniature speakers and microphones and to establish an arrangement similar in principle to the one used in the Concert Studio itself.

Different sizes and shapes of the interior ceiling 'waves' were modelled (in plastilina) and mounted in the ceiling of the model. The experiments suggested that a minimum width of the ceiling waves should correspond to the frequency of the mode in question, which meant that the ceiling waves in full size would have to be at least 4 m wide. This shape with 4 m wide waves had some smaller waves superimposed on it so that the final design had a more sophisticated appearance (Fig. 1.6). Although the model research showed conclusively that this ceiling design would have a sufficiently diffusing effect upon the pronounced modes it was considered wise to provide an extra possibility of reducing standing waves between the inner ceiling and the floor of the Concert Studio.

A large number of holes were distributed over the inner ceiling with a size which was large enough (6·5 cm) to act as apertures for Helmholtz resonators. The cavities behind the apertures were to be boxes made in concrete of suitable dimensions. These resonators could be tuned to have their absorption maximum at the frequencies of any insufficiently damped natural modes. In the empty studio the aforementioned test was repeated, and the results indicated that two modes, still detectable, were further suppressed after installation of the Helmholtz resonators. Resulting from the shape of this studio another acoustic problem could be expected: the focusing effect of the rear walls, which had their centre of curvature inside the studio, actually at the front part of the stage.

Preliminary investigations in the form of oscillogrammes of sound pulses, in the empty studio with untreated walls, showed very pronounced echoes, but even when covering the rear walls with massive layers of mineral wool a residual echo was still present. The finish was to be thin (3 mm) plywood with the highest possible perforation (25%). By arranging the mineral wool behind this panelling in narrow strips the absorption coefficient at high frequencies could be increased to 95% which improved conditions and reduced the echo. With this treatment the residual echo effect could be ascribed, not to reflections from the wall, but to reflections from the front side of the balcony (immediately above the wall). Even if this balcony front was bulging (and thus diffusing vertically) the curvature in the horizontal plane was enough to create a faint echo.

With the elimination of these anticipated defects before the completion of the auditorium, it was not expected that major problems would arise at a later stage. Indeed, the opening season (1945–46) indicated no such problems. The acoustics of the Concert Studio were generally acclaimed as being highly satisfactory, not only for transmissions but also for direct listening in the auditorium. The overall value of RT (1·5 s) was not impressive but there were no complaints about lack of reverberance. The RT v. frequency curve was practically flat from 100 to 2000 Hz (Fig. 1.8).

After a few seasons, some of the more advanced of the musical experts uttered their concern about the acoustics of the studio (to name only one

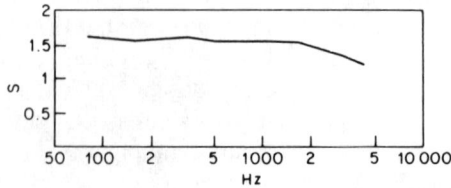

FIG. 1.8.   RT v. frequency curve—Concert Studio.

of these judges, a man of some eminence: the late conductor Erik Tuxen).

Tuxen was especially concerned about two facts: at the first (lower) balcony he had the impression of 'heaviness' of the orchestral sound, especially in tuttis. Also, when conducting, he experienced difficulties in hearing some of the weaker instruments such as the violins and flutes.

At about the same time, some of the members of the orchestra also began to utter criticism. In essence their remarks corresponded well with the second observation of Tuxen: difficulties in mutual hearing of some of the weaker groups (violins, flutes and oboes).

During the years 1951–55 several experiments were carried out with the aim of finding remedies for these acoustical deficiencies. (It should be mentioned that these problems did not affect the well established reputation of the studio as a remarkably good broadcasting studio.)

The two phenomena, the 'heavy' sound in the first balcony rows and the lack of 'communicating' sound on the platform, were apparently related and could be interpreted as being due to a common cause: too many of the early reflections were being directed towards the audience, away from the stage, by means of the fan shaped walls; and also, maybe, due to the sloping ceiling above the orchestra and to the considerable height of the ceiling over the stage area. By applying a series of slightly concave shaped reflectors arranged in a stepped fashion along the side walls, individually orientated parallel to each other (and to the axis of the hall), an attempt was made to reflect more sound back towards the musicians, but these attempts (according to members of the orchestra) were on the whole unsuccessful.

An entirely different approach followed: series of horizontally orientated reflector plates were suspended above the orchestra area. The reaction from the musicians was immediate and unanimous. This arrangement had apparently solved the mutual hearing problems. How was it possible to account for the effect of the reflectors by objective measurements? The story is told elsewhere but the main points are recounted here.

Series of noise pulses emitted from a point source on the stage were picked up by a microphone (located at different positions on the stage as well as in the audience area). The length of the noise pulses was varied from a few milliseconds to half a second. A high speed level meter indicated the sound level in decibels (dB) below the stationary level (the level corresponding to the longest pulse). A graph with corresponding values of pulse length (x-axis) and level (y-axis) illustrated the 'building-up curve' of the sound energy for the various locations of noise source and microphone (Figs. 1.9 and 1.10). Also indicated in the figures is the building-up curve (the fully drawn curve) to be expected if the process was exactly

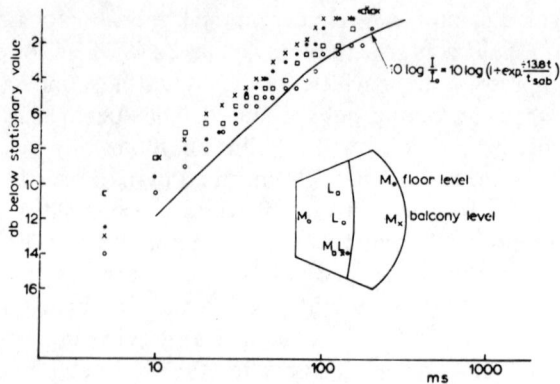

FIG. 1.9. Building-up process of sound energy—Concert Studio. Microphone (M) and loudspeaker (L) positions are indicated. The fully drawn curve represents the building-up process assuming it to be complementary to the decay process according to the equation given. The squares correspond to the measurements at the positions marked $M_\square$ and $L_\square$ and so forth.

complementary to the decay process for an RT value corresponding to that measured in the Concert Studio. The figures disclose that for certain locations on the stage of sound source and microphone the actual building-up curve is lagging compared with the calculated curve. They also indicate that the presence of the reflectors above the stage changes this situation markedly.

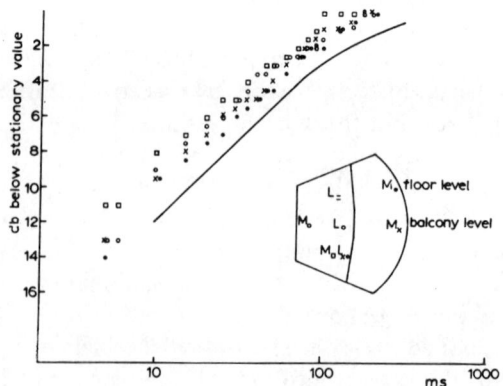

FIG. 1.10. Building-up process of sound energy—Concert Studio. Arrangement as for Fig. 1.9 but with glass plates suspended over the stage.

If the *Rise Time* (*TR*) is defined as the length (in milliseconds) of a pulse sufficient to reach a level of -3 dB, then some locations on the stage are shown to have *larger* values of Rise Time than locations in the audience area (and also larger than the corresponding calculated values). This situation has become known as *Inversion*. When the reflectors were in position, the values of Rise Time measured at the stage locations decreased and became smaller than the values measured in the audience area.

Assuming that the building-up process of sound energy in a room is complementary to the decay process and follows the same (reversed) exponential curve (in accordance with the simple Sabine theory), then numerical values of Rise Time may easily be calculated. It is easy to show (see Appendix I) that the theoretical value of Rise Time is very close to 5 % of the Reverberation Time. Calculated values of Rise Time may serve as a guide when comparing with values actually measured in a concert hall. Values in the stage area should be lower than values in the audience area and all measured values should preferably be lower than (or equal to) the theoretical value.

This kind of measurement and the application of the assessment of measured values is considered a new feature in experimental room acoustics. A musical criterion (Rise Time, TR) is measured in the stage area and in the audience area, and the average values for the two areas may then be compared by calculating the *Inversion Index:*

$$\text{II} \sim \frac{\text{TR (audience area)}}{\text{TR (stage area)}}$$

This index should preferably be equal to or larger than 1.0. No intricate theoretical considerations are offered for this working hypothesis, but merely a straightforward reasoning. The blending of the sound from the individual musical instruments will take place immediately in the platform area. If this blending does not occur, then the resulting sound field experienced by the audience will not be satisfactory for listening purposes.

This working hypothesis was suggested by the author, publicly, close to 20 years ago (at the 3rd ICA Congress in Stuttgart), but it would be an exaggeration to say that it has had very much response in professional circles (yet). However, it must be pointed out, that practising in accordance with this thinking over the years (since the early fifties) has assisted the author in solving many of the problems encountered in the design of concert halls and theatres. After the experiments conducted in the Concert Studio a permanent arrangement of horizontal reflectors above the platform was installed, at a height of 7–8 m above the platform floor and covering 40–45 % of the total area of their extension. This was done to the lasting satisfaction of the orchestra members and it did not impair the quality of the studio for broadcasting.

The other problem, that of 'heaviness' of the sound (as experienced at the first balcony) may probably be less noticeable after the installation of the reflectors.

It must be admitted that a complete solution of all problems inherent in the shape of this auditorium could hardly be expected from this one change, i.e. of putting reflectors above the stage. However, a significant confirmation of the importance of the reflectors was obtained some years later when they were taken away. The orchestra strongly demanded to have them back.

In the first design the reflectors were made of a special kind of glass which was supposed to be transformed to dust in the event of any breakage (for safety reasons). However, one night one of the reflectors fell to the floor in solid, large pieces heavy enough to kill, or seriously harm, anyone seated beneath. Next morning the whole set of reflectors was taken down and stored. But what happened? The musicians started complaining and a fresh solution had to be found.

Essentially, this meant changing the material from glass to plexiglass, which the fire authorities had refused to permit in the original design. A second edition of the reflectors (hexagonal this time) is still in use in the studio while these lines are being written.

Recently, speculation has started as to whether a major renovation could make the Concert Studio more adaptable for television purposes. It would hardly be appropriate to make such adaptations without seriously considering also the acoustical aspects. It would seem more natural to maintain the symphonic use as the prime purpose of the Concert Studio and to let any considerations of redesign start from that position.

The Tivoli Concert Hall, Copenhagen

Towards the end of the German occupation of Denmark several prominent buildings were bombed and destroyed in reply to the sabotage activities of the Resistance Movement. The old Concert Hall of Tivoli suffered this fate, and for a number of years after the war the popular Symphony Orchestra of Tivoli had to play in a smaller hall with unsatisfactory acoustics.

In the early fifties, however, the project for a new concert hall was underway, and the design of this hall was influenced by acoustical considerations from the outset. Both exterior and interior style had to be closely associated with the classic 'Tivoli' style, difficult to define, but well known to every Tivoli guest. Visual connection between the interior of the hall and the surrounding gardens was very important to the architects (Hans Hansen & Fritz Schlegel), and had to be established in spite of the difficulties this created for proper sound isolation.

So the application of glazed facades throughout the perimeter (on three sides) meant that it was necessary to have intervening spaces (entrance lobbies) which also were glazed towards the interior of the hall.

At the upper part of the side walls, there were to be windows situated directly between hall and exterior, and even if they were to be double pane with a large distance between the panes the resulting sound reduction would not be more than 40–45 dB. This meant that one had to tolerate intruding noises, the most dominant source of which were screaming youngsters enjoying the nearby 'hill and dale' course. It was agreed that this intrusion of background noise would be acceptable (even desirable) as part of the 'Tivoli environment'. The scope of the Tivoli Hall was defined as: 1, concert hall for symphonic music and 2, stage performances of ballet, drama and entertainment programmes. This meant that a complete fly tower with stage facilities and also an orchestra pit had to be included in the project. A seating capacity of 1600–1700 seats was the design brief. The layout envisaged a hall of moderate fan shape in plan with a largely uniform ceiling height having the high stage house as an extension and with two seating levels (floor, side and rear balconies). This layout presented a number of acoustical problems:

How to accommodate the orchestra?
How to avoid problems arising from the fan shape?
How to provide enough reverberance with a limited ceiling height (limitations due to appearance and economy)?
What shape to give the ceiling?

Experience from the Concert Studio indicated how important it was to have plenty of reflecting surfaces in the immediate vicinity of the orchestra, and the design became very much influenced by this thinking (Figs. 1.11 and 1.12).

The orchestra pit, arranged with a hydraulic lift, was to become the front part of the platform. The platform was to continue into the stage area, at both sides surrounded by vertical (stepped) panelling. These vertical panels were to be lifted into the fly tower whenever there was a stage performance. Above the platform area, rows of slightly tilted panels created a reflecting

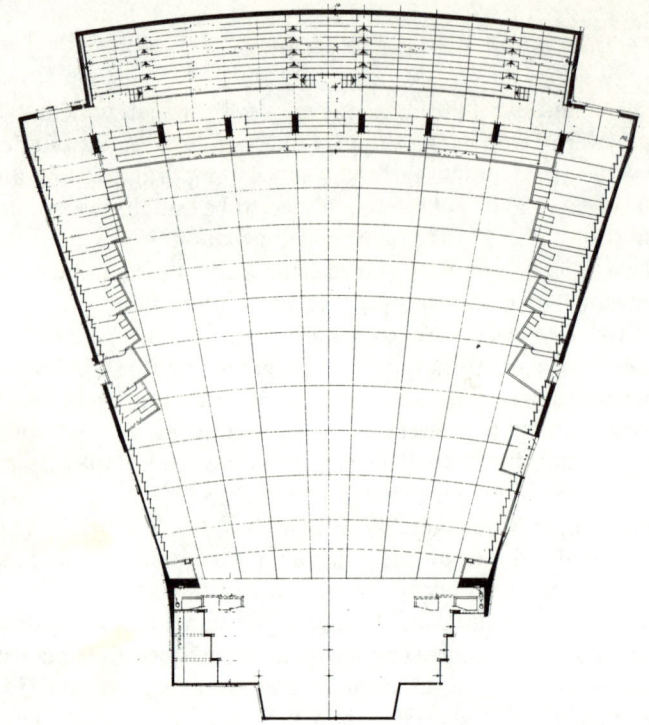

FIG. 1.11. Tivoli Concert Hall—plan.

FIG. 1.12. Tivoli Concert Hall—cross section.

ceiling (which alternatively might be opened in sections by automatically turning the panels vertical, for stage performances).

Towards the rear of the platform the reflecting ceiling was cut off so that the rearmost part of the platform was open towards the stage tower. The rear wall behind the platform area was to be a soft velvet curtain. It is a general experience that the heavy instruments (brass and percussion) do not need to have much reflected sound and that the absorbing surfaces improve the overall balance of the orchestra. The reflecting ceiling (in concrete and of considerable extent) above the platform area projects into the auditorium as a fixed reflector. The complete 'acoustic profile' of the stage and the hall is shown in Fig. 1.13.

FIG. 1.13. Acoustic profile of the Tivoli Hall.

The design of the auditorium ceiling may now be considered to be somewhat outdated. It applied the concept of reflecting the 'sound rays' off the ceiling, aiming to achieve an evenly distributed sound over the audience area, by zig-zag shapes of different orientation.

A more up-to-date concept of the acoustical function of the ceiling in a concert hall is that all reflections from above shall be as much diffused as possible. In the medium and low frequency range a ceiling shaped according to the above description will in fact be quite diffusing.

The undesirable acoustics which would be generated by the fan shape in plan is partly compensated by a special arrangement of stepped panelling along the side walls. The individual steps (width: 60 cm) are orientated to be parallel to the long axis of the hall. The reflections off these panels are diffuse at low frequencies but specular at high frequencies. The transition from diffuse to specular reflection occurs in the frequency region 500–1000 Hz, so that the important upper range of the orchestra sound spectrum has the advantage of being reflected off apparently parallel side walls. This design may be mentioned as an early example of the desire for 'side reflections', which has now been generally acknowledged as an important factor.

The limitations in the volume of the hall (approximately $7 \cdot 5 \, m^3/seat$) indicated that the RT for a capacity audience would be rather low.

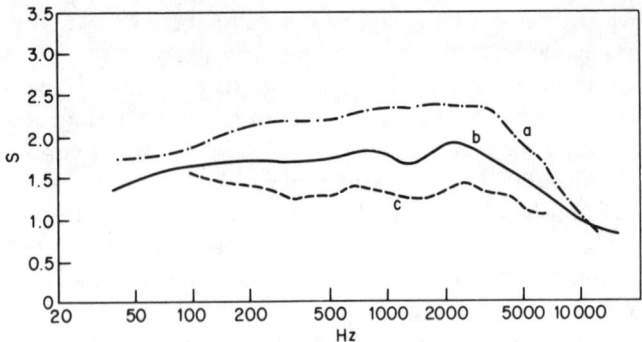

FIG. 1.14.   Tivoli Concert Hall—RT v. frequency curves. RT: a, empty hall; b, 70% capacity audience; c, capacity audience.

Therefore it was recommended that a light upholstered chair should be chosen and that there should be a certain variation in RT with the number of people attending a performance. Measured RT v. frequency curves are shown in Fig. 1.14.

Special precaution was taken to avoid the effect of the deep balcony on the hearing conditions for the rear-most rows. An overhead passage allows reflected sound to enter the rear area of the balcony from above. Although the ratio of height to width of this hall is in no way outstanding, the arrangement of the orchestra and the stage surroundings has contributed to the musical quality in a positive manner. Views of the completed auditorium are shown in Figs. 1.15 and 1.16.

The appraisal of the acoustical quality at the opening concerts was unanimous, and the rather low value of RT did not influence the

FIG. 1.15.   Tivoli Concert Hall—view towards the stage.

Fig. 1.16.   Tivoli Concert Hall—view across the hall.

assessments. No complaints were heard from the orchestra about mutual hearing problems of the various groups.

Obviously, it would be important to repeat the same kind of measurements of Rise Time as had been carried out in the Concert Studio. The result of such measurements showed that the values of Rise Time in the stage area were considerably lower than the corresponding theoretical values, whereas values measured in the audience area were comparable with, although still less than, the theoretical values. (Fig. 1.17.)

Contrary to what happened in the Concert Studio no major defects were ever subsequently discovered in the Tivoli Concert Hall. Of course, problems occurred when modern groups (rock, beat, etc.) eventually occupied the hall on some occasions. These problems seem to be common to all genuine concert halls and they may be explained by the difference in acoustical conditions actually required by the excessively loud sound produced by such groups and also by the strong emphasis on rhythmic elements, which do not require reverberance.

The experience gathered from these two halls has acted as a guide for the author in his approach in general to the design of concert halls and theatres. It also whetted his appetite for gaining a more thorough understanding of the development through European history of these two different concepts of auditoria design. These studies led to the realisation that in our century there has occurred a fatal tendency among architects to confuse what were

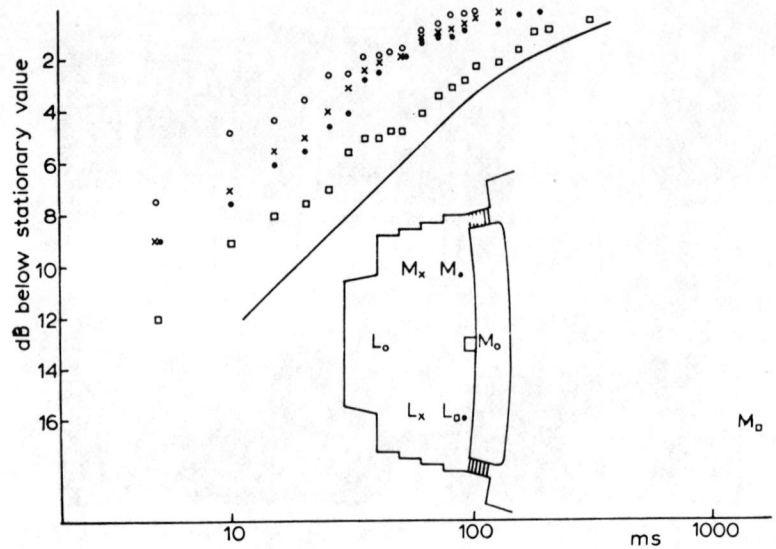

FIG. 1.17.   Tivoli Concert Hall—examples of build-up curves. Points marked
○, × and ☐ correspond to measured values; the fully drawn curve represents
the theoretical values.

originally two altogether different concepts and this confusion has already
led to many serious mistakes.

A brief historical diversion may therefore be regarded as inevitable at this
point.

BIBLIOGRAPHY

Subject	*Reference*
Radiohuset, Copenhagen—general description	Lauritzen, V. *Arkitekten*, special issue, 1946, Akademisk Arkitectforening, Copenhagen.
Radiohuset, Copenhagen—acoustics	Jordan, V. L. *Arkitekten*, special issue, 1946, Akademisk Arkitectforening, Copenhagen.
Sound absorption of perforated panelling	Zeller, W. *Akustische Zeitschrift*, **3**, 1938, p. 32.

Sound absorption of perforated panelling	Jordan, V. L. Elektroakustiske Undersøgelser. Doctoral thesis, Royal Technical College of Copenhagen, 1941. Jordan, V. L. Akustische Zeitschrift, 5, 1940, p. 77.
RT v. frequency curves	von Békésy, G. Ann. d. Physik, 5, 1934, p. 665. Jordan, V. L. J.A.S.A., 19, 1947, p. 972.
Concert Studio, Radiohuset—suppression of natural modes, building-up process of sound pulses	Ibid Jordan, V. L. Proceedings, ICA III, Elsevier Publishing Co., Amsterdam, 1959, pp. 922–925.
Complementarity of build-up and decay processes	Schroeder, M. R. J.A.S.A., 40, 1966, p. 549.
Concert Hall of Tivoli, Copenhagen.	Jordan, V. L. Ingeniøren, 65, 1956, p. 861.

DATA FOR CONCERT STUDIO, RADIOHUSET, COPENHAGEN

Year of completion: 1945
Volume: 11 700 m³
Total area: 930 m²
Number of seats

Orchestra level:	380
1st balcony:	375
2nd balcony:	338
Total	1093

RT measured in empty hall (noise, 1/3 octave filter):

80	125	250	500	1 000	2 000	4 000	6 400	(Hz)
1·6	1·6	1·7	2·0	2·0	1·9	1·2	1·0	(s)

RT measured in hall with capacity audience, orchestra (84) and choir (94) (chords, 1/3 octave filter):

80	125	250	500	1000	2 000	4 000	6 400	(Hz)
1·6	1·6	1·6	1·5	1·5	1·5	1·2	—	(s)

General Description

side walls:	wood panels (16 mm), every second strip with air behind, the rest direct on solid wall
rear walls:	orchestra level, perforated plywood panel with mineral wool behind; balcony levels, wood panel without perforation
ceiling:	profiled wooden panelling plastered on concrete.
floors:	parquet
seats:	upholstered, leather
platform size:	220 m²

DATA FOR TIVOLI CONCERT HALL, COPENHAGEN

Year of completion: 1956
Volume: 12 700 m³
Total area: 1160 m²
Number of seats
   Orchestra level: 1308
   Balcony level: 468

   Total   1776

RT measured in empty hall (noise, 1/3 octave filter):

67	125	250	500	1 000	2 000	4 000	6 000	(Hz)
1·6	2·1	2·1	2·2	2·3	2·4	2·1	1·8	(s)

RT measured in hall with capacity audience (chords, 1/3 octave filter):

67	125	250	500	1 000	2 000	4 000	6 000	(Hz)
—	1·5	1·35	1·3	1·3	1·35	1·3	1·1	(s)

General Description

side walls:	wood panels (16 mm) with air space behind vertical strips of panelling parallel to long axis of hall
rear walls:	orchestra level, glass, inclined; balcony level, wood panels on concrete
ceiling:	gypsum, surface of hard fibre board
floors:	rubber tiles on concrete and on wood upon risers
stage walls:	wood panels (16 mm)

stage ceiling:	wood panels, horizontally for concert, vertical for ballet or comedy
seats:	upholstered seat and back, perforated wood underside of seat
pit size:	40 m^2
platform size:	130 m^2

Chapter 2

Brief Historical Survey

The classical concept of a theatre is the ancient Greek amphitheatre (e.g. Epidauros, Fig. 2.1), where the actors are located in a central area and the audience is seated on a slope with all seats (steps) orientated towards the actors' area, thus creating a semicircular or semielliptic plan. It is a straightforward solution when the aim is to have a capacity audience all on the same level, as close as possible to the stage, and to have optimum sound distribution and maximum speech articulation. The Greek theatre had very little reflected sound, merely that from the stone pavement of the central area and from a building just behind this area.

The Roman amphitheatre (e.g. Herodes Atticus at Athens, Fig. 2.2)

FIG. 2.1.   Greek Amphitheatre, Epidauros.

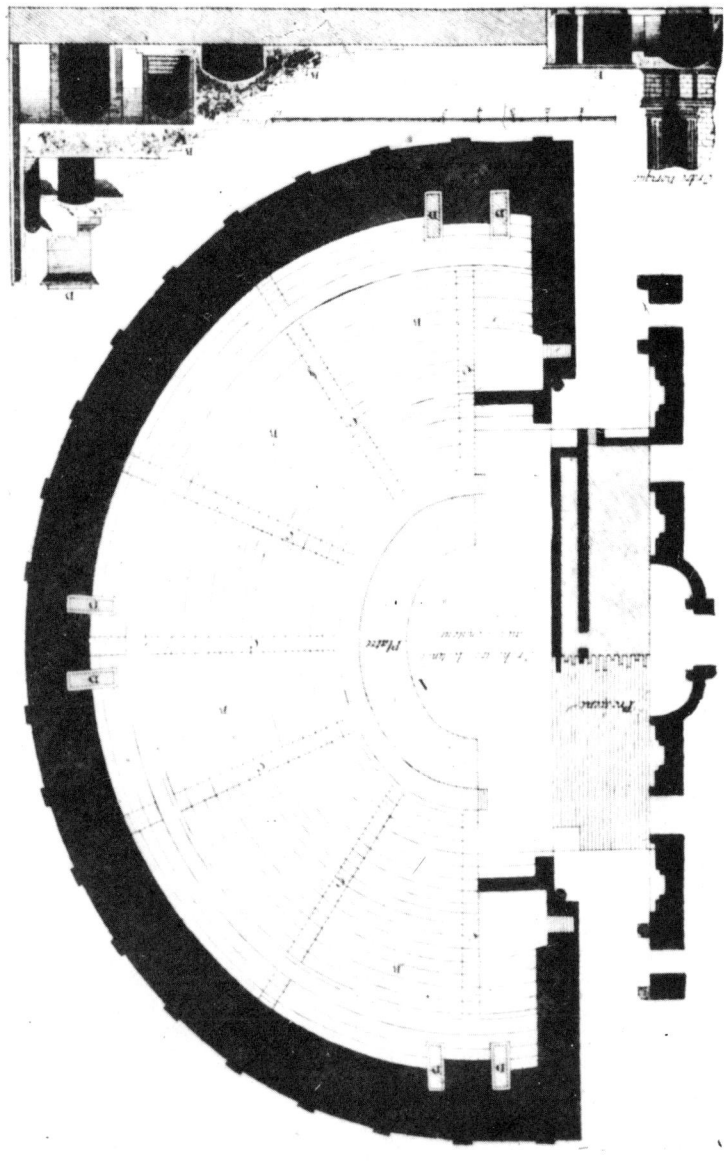

FIG. 2.2. Roman Amphitheatre, Herodes Atticus—plan and view.

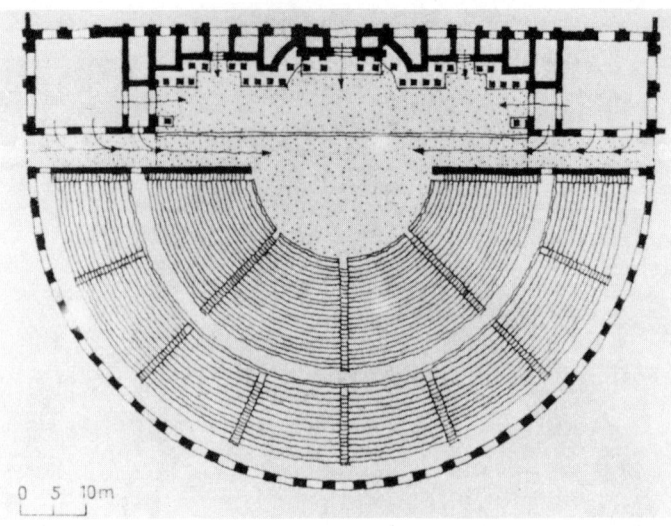

FIG. 2.3.   Roman Amphitheatre, Orange—plan.

FIG. 2.4.   Roman Amphitheatre, Orange—view.

developed a steeper seating ramp and a large building behind the acting area with more reflecting surfaces, some of which also reflected sound from the sides. All the reflections have short time delays and therefore essentially serve to increase the strength of the direct sound; the side reflections may also provide a certain 'space effect'. Another example of the Roman amphitheatre is the larger theatre of Orange in France (Figs. 2.3 and 2.4). The layout, with the elevated passage in front of the stage building, is the same as in Herodes Atticus.

A whole millenium passed before the theatre concept further developed during the Italian Renaissance. An early example of Renaissance theatre buildings is Teatro Olympico, Vicenza (Figs. 2.5 and 2.6). The transition

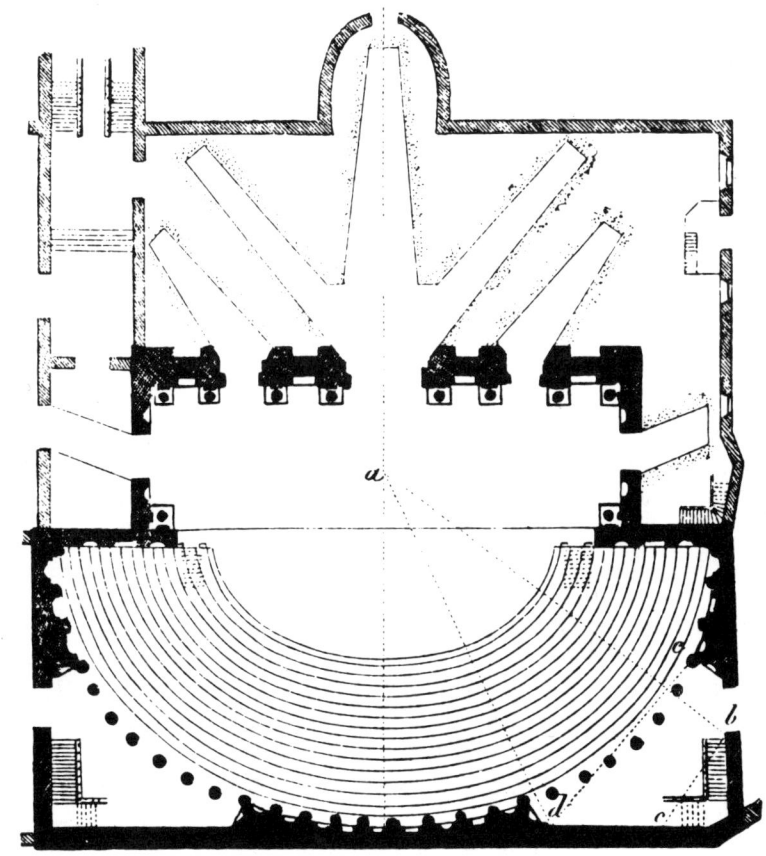

FIG. 2.5.   Teatro Olympico, Vicenza—plan.

FIG. 2.6.   Teatro Olympico, Vicenza—view.

from the open air theatre to the indoor theatre is here complete, but the
layout of stage and audience seating is still very close to the plan of the
Roman theatre.

For the first time in the history of the theatre we have here an enclosed
volume, which means that the sound now suffers repeated reflections and a
regular reverberation process, added to the direct sound and the early
reflections. Without an audience present this theatre must be rather
reverberant, even taking into account the many openings in the stage wall.
With an audience present it is possible that the RT is reduced to a value
which does not influence speech articulation too much, especially in view of
the abundance of early reflections; however, the RT at low frequencies must
dominate because of the many stone and marble surfaces.

Another important step in the development of the theatre is represented
by the Teatro Farnese in Parma (Figs. 2.7 and 2.8) which was built a few
years later. Teatro Olympico had a wide semielliptic seating area and was
quite shallow in depth, but Teatro Farnese had an oblong shape with a
seating area composed of a semicircular section at the rear and two straight
sections along the sides, leaving a large open space in the middle, linked to
the stage area. At the same time the stage developed into two distinct areas: a
forestage in front of a portal and a main stage behind the portal.

Acoustically, this indicates the beginning of the problematic proscenium
theatre where the stage house consists of a separate (acoustical) space,
coupled to the auditorium space through the proscenium opening.

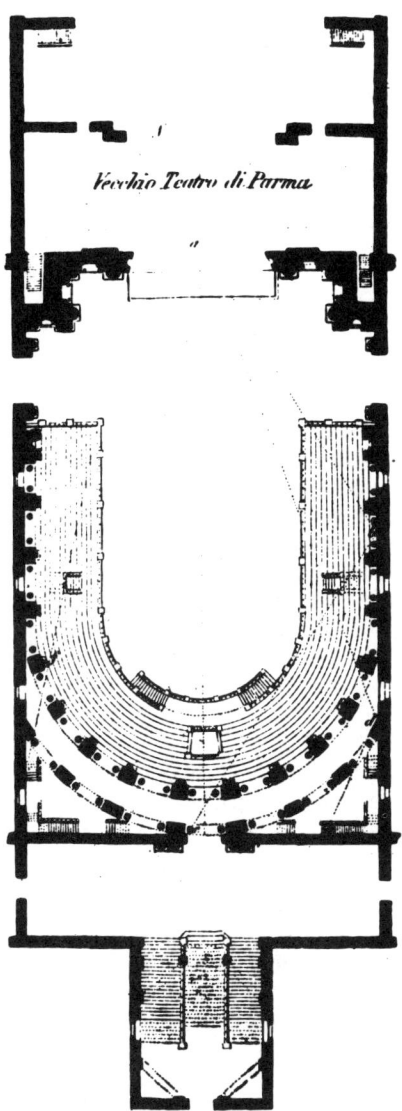

Fig. 2.7.   Teatro Farnese, Parma—plan.

FIG. 2.8.   Teatro Farnese, Parma—view.

Another important feature is the gradual development of a schism between actors (singers) on the stage (forestage or main stage) and musicians in an area in front of the stage. In the Greek amphitheatre a few actors, a choir and a few instrument players were all located more or less in the same acoustical environment, the actors using masks to reinforce their voices. In the Medieval and Renaissance theatres no clear distinction existed between performance areas for acting (singing, dancing) and for the musical accompaniment. The creation of the art of Opera, however, increased the size of the musical ensemble which gradually developed to a full sized orchestra, located in a separate area in front of the stage, which was lowered and became the orchestra pit. The schism is completed with the advent of the lyrical theatre a century later, e.g. Teatro San Carlo in Naples (Fig. 2.9).

Consider the contrast, acoustically, between a stage house where fly tower, decorations, carpets and props of all kinds constitute the surroundings of the singers, and a narrow pit with sound reflecting floor and walls, open towards the auditorium, which constitute the surroundings of the orchestra. No wonder that the problem of balance between singers and orchestra became a major acoustical problem in many of the world's opera houses.

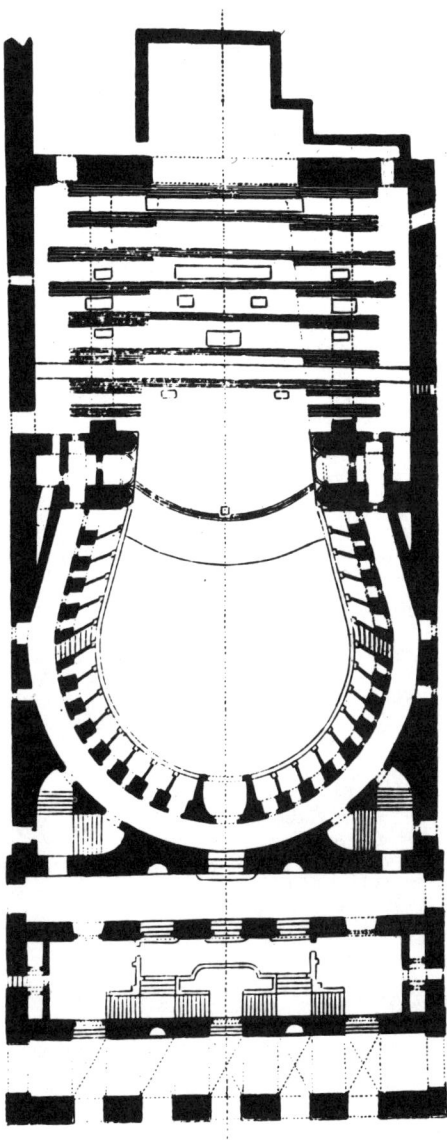

Fɪɢ. 2.9. Teatro San Carlo, Naples—plan.

Looking at a plan of Teatro San Carlo we notice the area between the stage proper and the orchestra pit. This area is actually a kind of forestage and acoustically it represents a transitional space between the stage house and the auditorium. It seems very probable that a large part of the important acting (and singing) takes place in this area where the singers will have the benefit of reflections from the sides.

The opera building in Milan, the old Teatro alla Scala had exactly the same feature as San Carlo, a deep proscenium arch which acted as a frame for the front part of the stage and for the orchestra pit. (When the Scala Opera was rebuilt after the war the stage was moved further back so that the orchestra pit now starts at the proscenium frame.)

Not all the old opera buildings have this feature, a good example being the old Metropolitan Opera of New York (Figs. 2.10 and 2.11) which had quite a shallow proscenium frame. Teatro Colon of Buenos Aires (inaugurated in 1908), on the other hand, preserved the Italian tradition; its design was actually almost copied from San Carlo of Naples. A view towards the stage in Teatro Colon (Fig. 2.12) shows very clearly one side of

Fig. 2.10.   Old Metropolitan, New York—view.

FIG. 2.11.   Old Metropolitan, New York—view.

FIG. 2.12.   Teatro Colon, Buenos Aires—view.

the proscenium frame (with boxes). Incidentally, this view also shows the arrangement used at symphony concerts, where the pit floor is elevated to the level of the stage and a reflecting shell construction is placed behind the rear part of the orchestra.

Some years ago a German acoustician (Fritz Winkler) collected opinions among a group of famous conductors on well known large concert halls. It so happened that an *opera house*, Teatro Colon, topped the list as the best *concert hall*. This, in itself, does not prove anything, but if one starts thinking in terms of side and ceiling reflections, it is easy to see what an abundance of side reflections must be present in Teatro Colon and how relatively unimportant must be the ceiling reflections in the same auditorium.

Most of the old opera auditoria have a lot of wood in their construction and wood panelling in the huge areas of loges, boxes and galleries. This means that low frequency absorption is taken good care of so that in most cases the RT v. frequency curve becomes fairly flat.

A noteworthy example of an opera building where the classic Italian tradition of the many tiers and balconies was abandoned is the Wagner

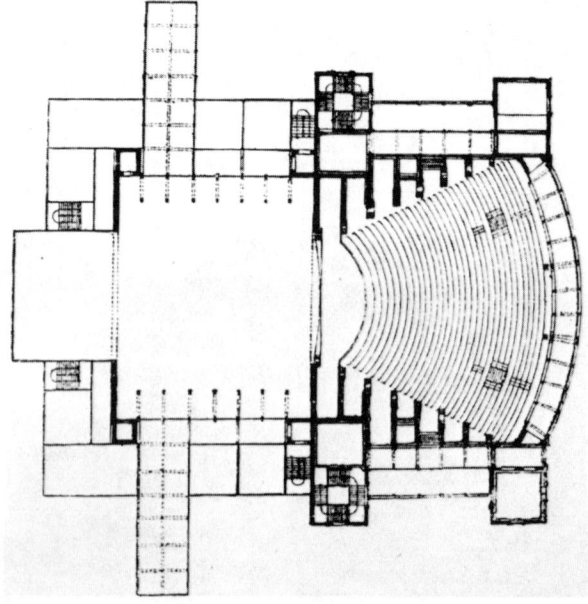

Fig. 2.13.   Wagner Festspielhaus, Bayreuth—plan.

FIG. 2.14.   Wagner Festspielhaus, Bayreuth—view.

Festspielhaus in Bayreuth (Figs. 2.13 and 2.14), which has been much acclaimed for its excellent acoustics. The audience seating is mostly at floor level, with a rather steep slope. The boundaries (ceilings and side walls) are essentially flat and the side walls are parallel. The appearance of a fan shaped auditorium is nevertheless achieved by a row of niches and pillars framing the seating area. This arrangement provides an extraordinarily strong diffusion of the side reflections, and it also means that the volume per seat (related to the main seating area) is larger than in most other opera house auditoria with the same capacity. Even when fully occupied this auditorium is quite reverberant (RT: 1·5–1·6 s).

Another feature which is unusual is the enclosed pit where the opening towards the auditorium is very limited. No doubt, this is advantageous for the balance between singers and orchestra.

The classical concept of a concert hall is much younger than that of a theatre. As the symphony orchestra has developed from smaller ensembles of the 18th century so the concert hall developed from the smaller recital hall. As the ordinary shape of most rooms and halls is, by convention, rectangular with a flat ceiling, so the music room, whether a recital hall or a concert hall, originally had the same shape. The prototype of later and larger concert halls could be said to be 'Altes Gewandhaus' in Leipzig, which was built in 1780. (It existed for more than a century and was succeeded by 'Neues Gewandhaus'.)

Altes Gewandhaus was a small hall, which in the beginning only seated 400, but it was later increased in seating capacity to accommodate nearly 600. The volume (probably 2000–2500 m^3) must have been undersized for a full orchestra of modern proportions but we may assume that the orchestra playing in Altes Gewandhaus probably included only 45–50 musicians.

In this hall the audience was seated in a peculiar manner. They sat in rows parallel to the long axis of the hall (supplemented at the rear with rows across), an interesting configuration which was hardly chosen for acoustical reasons, but maybe for social—several rows exclusively with ladies, confronting each other! (Fig. 2.15.)

FIG. 2.15.   Altes Gewandhaus, Leipzig.

'Neues Gewandhaus', built in 1886, had dimensions more in accordance with modern concepts of a concert hall. It had a total seating capacity of 1560, with two-thirds on floor level and the rest on two balcony levels. The lower balcony continued all the way round the perimeter, interrupted only by the organ which occupied a central position behind and above the platform. This concert hall (destroyed during the war) had predecessors,

FIG. 2.16.   St. Andrews, Glasgow—view.

e.g. Musikvereinsaal (Vienna, 1870), Stadt Casino (Basle, 1876), St Andrews (Glasgow, 1877, see Fig. 2.16), and successors, e.g. Concert-gebouw (Amsterdam, 1888) and Symphony Hall (Boston, 1900). They are all excellent examples of the distinctive group of the 19th century classical concert halls, which have many important features in common.

They are all rectangular, with high ceilings, one or two shallow balconies, richly ornamented and most of them with extended areas of wood panelling. Each of them has individual features as well, but their acoustical properties cannot be too different. Often, therefore, individual preferences between these halls may be based more on subjective than on objective observations.

Talking of the classical concert halls as a group with common features, it must be noted, however, that one physical parameter of the Boston Symphony Hall differs from the European halls—the size. Whereas the five European halls vary from between 1400 and 2200 seats, the Boston hall has seating for 2600 plus. The Boston hall was also the first hall ever built where mathematical predictions of reverberation time had been undertaken, actually by the very same scientist who first studied the reverberation process on a scientific basis and who defined the concept of RT (named after him: the Sabine reverberation time, i.e. the time interval in which the sound energy level diminished by 60 dB). By applying such mathematical predictions of RT, Sabine was able to make certain that the considerable increase in volume of the Boston hall compared with the Gewandhaus hall did not result in any undesirable increase of RT.

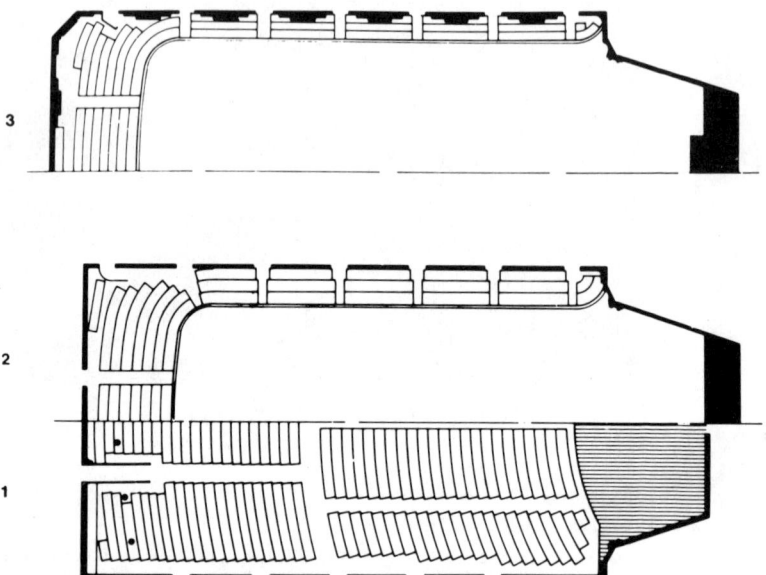

FIG. 2.17.   Boston Symphony Hall—plan. Seating capacity: (1) 1486; (2) 598 and (3) 547.

In designing the Boston Symphony Hall Sabine also applied a feature which was not present in any of the classical European halls (at least as explicitly as in the Boston hall). He placed the orchestra in an extension to the body of the hall, something very close to what may be called a separate enclosure, where ceiling and surrounding walls come closer to the orchestra than ceiling and walls of the main body of the hall (Figs. 2.17 and 2.18). Apparently he did this by instinct since nobody at that time, Sabine himself included, knew anything about the vital importance of the early reflections.

No doubt, it was a master stroke and the reputation of the Boston Symphony Hall—that it belongs to the very best of all the classical halls— may quite likely be intimately associated with this specific design feature. Looking at the development in design of concert halls in our own century we notice an increasing interest in experimenting with shape and materials. Salle Pleyel has already been mentioned but there are many other examples.

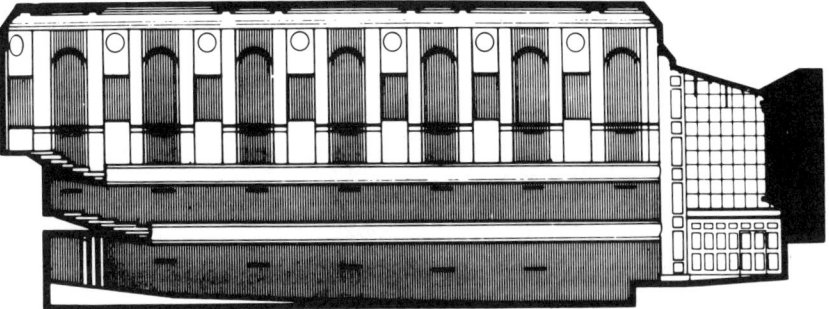

FIG. 2.18. Boston Symphony Hall—section.

We also note a trend towards confusing theatre and concert hall design. The visual experience becomes more and more important (reinforced by the modern technological arts of cinema and television) and this is apt to dominate the more subtle experience of hearing and listening to music.

We are still at the stage where we confuse the categories of theatrical and musical requirements and, as if this problem were not enough in itself, we are about to enter a new era in the design of large halls, an era where the same hall is expected to be suitable for several (sometimes contradictory) purposes: the era of the Multipurpose Hall.

BIBLIOGRAPHY

Subject	*Reference*
Development of the theatre	Graf, H. *Opera for the People*, University of Minnesota Press, 1951.
Different distribution of the audience	Cremer, L. In *Auditorium Acoustics* (Ed. R. McKenzie) Applied Science Publishers, 1975.

Opera auditoria, concert halls

Beranek, L. *Music, Acoustics & Architecture*, John Wiley and Sons, New York, 1962.

Conductors' opinion on halls

Winkler, F. *Bauwelt*, **48**, 1957, p. 1349.

Chapter 3

Multipurpose Halls—Early Period

Use of a large hall for more than one purpose is not unusual but it is still not so usual to plan and design a large hall such that it becomes particularly suited for a definite list of purposes. The use of the word 'list' is deliberate, since the important question to put to the potential client, user or architect, is this: 'What is the list of priorities?'

No large hall can ever be equally suitable for many different purposes. Given a definite list of priorities, however, the acoustic consultant will be able to adapt the design of the hall to any particular situation and to aim at a superior solution for the most important purpose without neglecting the subsequent items on the list, but weighing them according to their position. In many cases musical purposes will rank high on the list and also be of special concern. When raising questions about the different musical purposes—symphonic music, opera, musicals, operetta etc.—they would have to be stated in the appropriate order. Other purposes such as assembly, congress and speech have requirements which do not coincide with the musical purposes but which also definitely imply acoustical considerations.

When working on the acoustical design of multipurpose halls it is necessary to define the design possibilities and to emphasise the choice between the two different main approaches: working with the various factors of *natural* acoustics of a hall means working with volume, shape and surfaces (reflecting, absorbing, diffusing), possibly also with mechanical changes of the character of some of the surfaces; and working with *artificial* acoustics means working with microphones and amplifiers plus several electrical artifices like time delay and artificial reverberation, and with different arrangements of loudspeakers.

39

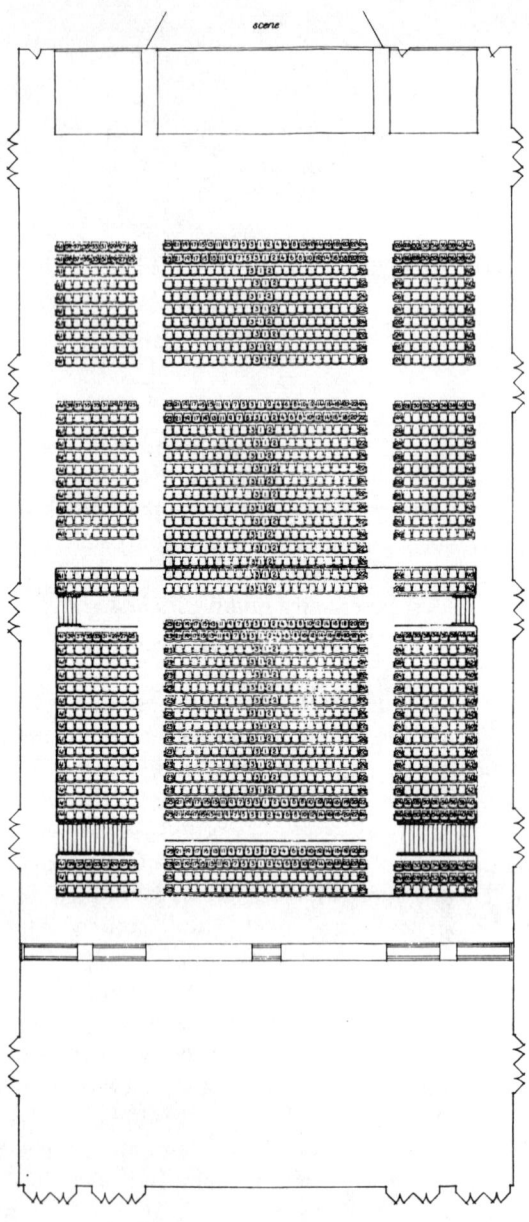

FIG. 3.1.   Aalborghallen—plan.

In practice, there may not be a clear division between these two approaches since the design of many multipurpose halls will have to rely on both to achieve optimum results. Nevertheless, it is important that the acoustician decides in advance which of the two approaches will have the lead, because this decision will influence the subsequent detailed design. Some early experiences of the design of multipurpose and dual-purpose halls, namely: 1, Aalborghallen; 2, Scala Theatre (of Aarhus) and 3, University Hall (of Reykjavik), will now be described. Whereas Aalborghallen involved a real multipurpose situation, the other two halls had only two or three purposes.

Aalborghallen

The municipality of Aalborg had already issued, during the war, the conditions for an architectural competition, to provide a new cultural centre for the city. The winning team (architects Preben Hansen, Otto Frankild and Arne Kjaer) was commissioned to prepare a detailed plan; the actual construction started in 1949 and the building was completed by the end of 1952. The complex consists of a large hall surrounded on two sides by lobby, foyer and a number of smaller halls. In the following paragraphs only the large hall is described. Various aspects of the hall and its design are shown in Figs. 3.1, 3.2, 3.3, 3.4 and 3.5.

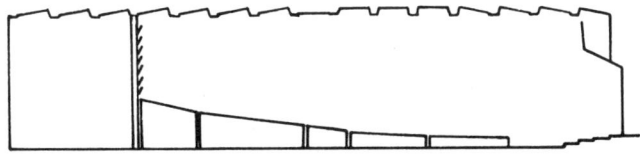

FIG. 3.2.   Aalborghallen—longitudinal section.

At the top of the priority list was an assembly hall (with more than 3200 seats) but the following two items (i.e. proposed uses) on the list were very close to this in importance: a concert hall (with 1800 seats) and an opera/drama theatre (with 1400 seats). Other items were: banquets, circus performances and variety shows. The architects had wisely concluded that the only sensible approach to this multipurpose situation was to use a basically rectangular shape and to adapt the length of the hall according to each specific capacity and purpose.

This adaptation was to be realised by a large, movable partition, moving in the direction of the longitudinal axis of the hall, thus dividing it into two smaller sections. The section closest to the stage (a full theatre stage with fly

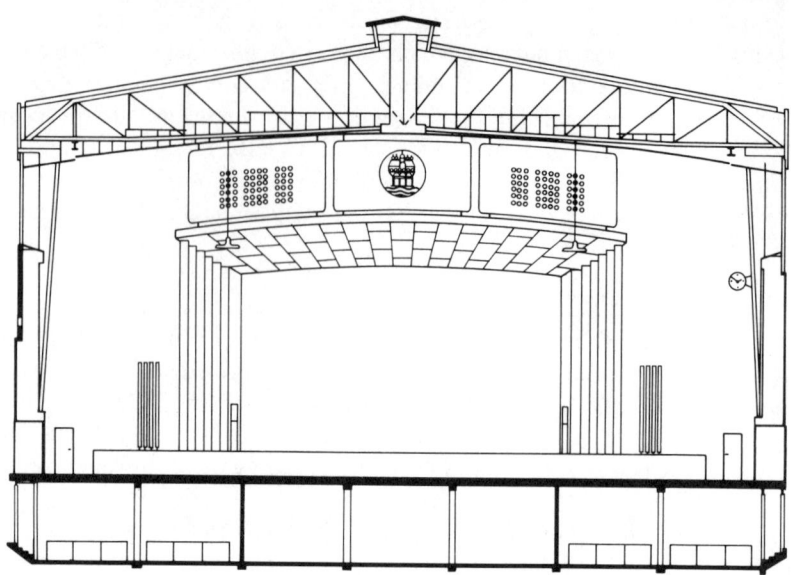

FIG. 3.3.   Aalborghallen—cross section.

FIG. 3.4.   Aalborghallen—view towards stage.

Fig. 3.5. Aalborghallen—view towards rear.

tower) became either the concert hall or the opera/drama theatre. The other section could be simultaneously available for other purposes (e.g., for a social gathering or meeting) and this meant that the serious problem of sound isolation of the movable partition had to be solved.

A sound reduction figure of at least 55–60 dB had to be achieved, and this was only feasible by having two completely independent sections of the movable partition, each being 35 cm thick and separated by a distance of 30 cm.

Since the hall had a number of equally spaced, load carrying columns and beams, these could act as frames for the partitions to meet, the slots between frame and partition being of the order of only 10–15 mm and fitted with sound absorbing lining along the perimeter of the wall. Each section of the partition had a multiple sandwich construction and had a highly efficient sound absorbing treatment exposed on the inner surfaces facing the intervening space. Sound reduction measurements were taken when the hall and the partition were completed and they showed that the design goal of sound reduction of 55–60 dB had been obtained in practice.

The design of the hall has many features which are based upon acoustical considerations. First of all, the seating on a flat floor is supplemented with a section of seats at the rear, raised in the manner of an amphitheatre. This section is mounted on waggons which may be rolled along to the orchestra lift and taken to the basement for storage under the main stage.

The ceiling has panel sections (between the beams) which are angled differently according to their position, the aim being the same as in the Tivoli Hall: to create an even distribution of ceiling reflections, so here, also, the idea is a little outdated. The side walls are also panelled (between the columns) but here the panelling is shaped in triangular sections, to diffuse the side reflections.

The rear wall (movable) has a large number of single panels protruding from the wall and angled to reflect the sound downwards to the rear rows of seats.

Since the volume of the hall is large compared to the number of seats (approximately $14\,m^3$/seat) it was necessary to provide some additional absorption. This was achieved by applying perforation to some areas of panelling.

By using 16 mm plywood and perforating the plywood with circular holes the absorption became selective and was calculated to absorb in the medium frequency range (400–1000 Hz) deliberately to create an RT v. frequency dependence similar to the preferred characteristics which had been found by experimentation at Radiohuset in Copenhagen. After completion, measurements were made in the concert hall (with a capacity audience) and the results indicated that the intention had been achieved (see Fig. 3.6); i.e. a slight increase of RT in the frequency range 2000–4000 Hz and also a slight increase of RT in the low frequency range.

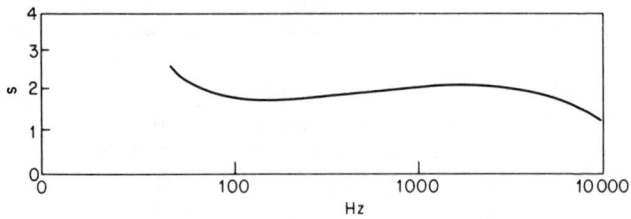

FIG. 3.6.   Aalborghallen—RT v. frequency (Concert Hall with capacity audience).

The arrangement of the orchestra also needs some comment. The entire platform is situated in front of the main stage curtain, which is in a lowered position when the orchestra is playing at a concert. The platform is surrounded on both sides by two large (hinged) side reflectors, and there is also an overhead reflector in front of the proscenium. These reflectors serve the same purpose as the orchestra enclosure of the Tivoli Hall, providing early reflections for the musicians.

Measurements of rise time have never been taken in Aalborghallen, since the method of measurement was developed after the completion of this hall, but there are reasons to believe that no inversion phenomena will occur in Aalborghallen.

The opera/drama theatre (1400 seats) is established by moving the partition two sections closer to the stage. Reverberation time is not reduced very much by this operation so it stays rather high, around 1·8 s, which is a high value for opera and certainly too high for drama. Admittedly, here is a case where the concert hall situation has been given priority over the drama theatre.

What about the assembly hall, with more than 3000 seats? The RT does not increase much when moving the partition flush to the solid rear wall, but for speech this situation undoubtedly calls for a public address system of high quality. Due to limited experience with correct placement and proper type of loudspeakers the original solution was not satisfactory—on the contrary. The two main speaker systems were identical, each consisting of a huge baffle with 12 speakers arranged in a circle (the idea being that the beaming effect of a linear array would be thus modified into a focusing effect, the focal area being the seating area in the rear part of the hall). This arrangement may to some extent have worked the way it was intended, but the fundamental error (as it was soon discovered) was the location of the speaker systems at a high level—on top of the proscenium arch, slightly angled downwards. This meant that there was a complete lack of coverage of the seats in the front and the middle of the hall. Very soon the proscenium speaker system was supplemented with a pair of column speakers (1·5 m length) placed at both sides of the stage (at stage level). This was a definite improvement but still did not provide a satisfactory coverage over the whole of the audience area, which in the case of the assembly hall is 70 m long.

A period of experimentation followed. By using a number of additional column speakers along the side walls the coverage was improved and this became the final solution back in 1953. There were still problems with the time delay between the speakers at the rostrum and those along the side walls, especially in the rear of the hall. The systems for artificially delaying the loudspeakers (to achieve a natural impression) were not developed at the time and were not commercially available.

In connection with a recent redecoration of the hall (1975–76) the loudspeaker systems have been revised and delayed speech has been introduced.

Several opening events of Aalborghallen also included a preliminary test of the large assembly hall used as a large concert hall. For this event none other than the world-famous Italian tenor, Beniamino Gigli, was invited to let his powerful voice fill the hall (which indeed it did). This test turned out to be a good omen for the later official opening, early in 1954. The enthusiasm for the acoustics on that occasion confirmed the earlier experiences.

Later, of course, came criticism, not of musical events in the concert hall, which received only favourable comments, but of dramatic performances in the opera/drama theatre which were criticised for lack of articulation. This was understandable because of the large value of RT, and it only stressed the fact that a multipurpose situation tends to treat the different purposes with different expediency. Preferably, this variation should be in accordance with the list of priorities as established when the programme for the hall is decided.

Accepting the shortcomings as a drama theatre, it may be said that most other purposes have reasonably good conditions in Aalborghallen.

Scala Theatre, Aarhus

The Scala Theatre of Aarhus (built in 1953) serves quite a different function from Aalborghallen. It had only to serve two purposes: 1, concert hall, and 2, cinema (!).

The building site was limited in size and was of triangular shape; in fact, an empty lot behind the old City Theatre of Aarhus. The building is illustrated in Figs. 3.7, 3.8, 3.9 and 3.10. There were also limitations in building height, due to the City code for buildings in that particular area, and these limitations in height actually dictated the shape of the exterior

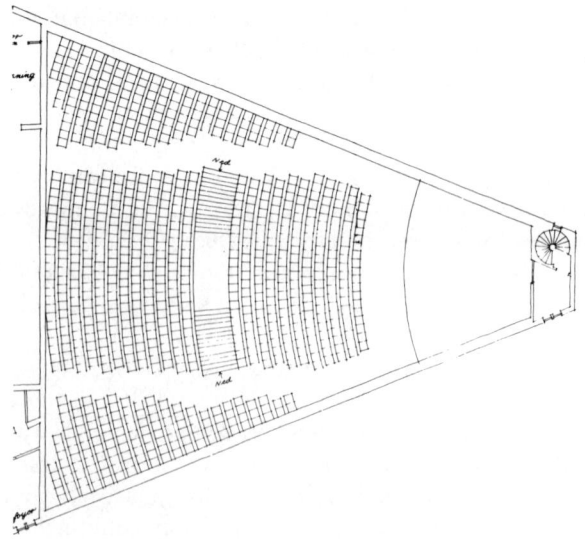

FIG. 3.7.   Scala Theatre, Aarhus—plan.

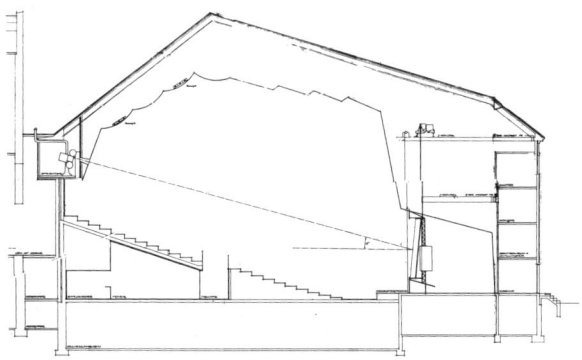

FIG. 3.8.   Scala Theatre, Aarhus—longitudinal section.

roof and thereby also the interior ceiling shape which had to be steeply sloping in cross section.

The seating also became restricted to a rather low figure (800 seats). On top of all these limitations the secondary purpose, i.e., use as a cinema, imposed rigid requirements on the slope of the floor to allow satisfactory sight lines.

FIG. 3.9.   Scala Theatre, Aarhus—view towards stage.

FIG. 3.10. Scala Theatre, Aarhus—view across.

All the restrictions meant that the volume per seat became rather low (approximately $6\,m^3/seat$). The predicted value of RT for the concert hall would not exceed 1·5 s. This, of course, was too high a value for the cinema. For a good cinema with a wide screen system, the value of RT should not exceed 1·0 s. In these circumstances it seemed inevitable to introduce a system of variable acoustics. Certain experiences with such systems existed from some broadcasting studios completed at that time, in an extension to Radiohuset. The principle applied was rather simple: long strips of carpet material, placed along the walls (under open wood grills) and which could be mechanically rolled together and stored above the suspended ceiling. Behind the carpet strips was bare concrete which meant that the variation in absorption coefficient could be as much as 50–60 %. The area to be treated included both side walls and a part of the rear wall.

The wooden grill in front of the carpet strips was designed so that the walls looked very much the same whether the carpet strips were exposed or not, so that the system was architecturally acceptable.

After completion, measurements of RT indicated that the total variation obtained, in the medium and high frequency range, was 0·4 s, i.e., from 1·1 s for the cinema to 1·5 s for the concert hall. This was acceptable, but by no means an exceptionally good result.

The shape of the auditorium was triangular with a narrow stage and with a low ceiling above the stage. The low ceiling was required to accommodate the cinema screen in the space above the stage. To reduce the impact of the heavy instruments some slots were provided in the stage ceiling. This was only a modest attempt to provide a better balance for the orchestral music, an attempt which would have much less impact than the provision for the stage at the Tivoli Concert Hall, i.e., the omission of the ceiling reflectors at the rear of the orchestra.

Even before the opening concert, strong doubts about the acoustical quality of the hall were expressed, and it soon became apparent that one of the serious drawbacks of this hall was the narrowness of the stage together with the steep rake of the seating. This rake started very close to the platform, so that the direct sound from the strings got lost almost immediately, due to absorption by the audience in the first rows.

It was very effectively demonstrated during the opening performance that this explanation was valid. The orchestra played the national anthem with the audience standing which made the loss of string tone even more noticeable. The obvious remedy, to increase the level of the platform, was soon proposed and adopted. This improved the conditions somewhat, but did not solve the problem.

To make a long story short: serious limitations of volume and shape more or less dictated the acoustical conditions of the completed auditorium which also suffered from the specific dual purpose situation.

University Hall, Reykjavik

The Reykjavik University Hall (built in 1962) is a good example of a hall where the same two purposes, so much in opposition, of concert hall and cinema were asked for, and where a better solution was achieved.

The building site was of ample size and the dimensions and proportions of the hall were left to the discretion of the architect (Gunnlaugur Halldorsson) and his acoustical adviser. The building is illustrated in Figs. 3.11, 3.12, 3.13 and 3.14.

The list of priorities was: 1, assembly hall for the University; 2, concert hall for the Reykjavik Symphony Orchestra; 3, cinema, and 4, stage shows.

The more spacious layout made a more gentle slope of the seating feasible. With a seating capacity of 1000 seats a good public address system had to provide speech facilities for the assembly hall so that the secondary purpose, use as a concert hall, became the main target for the acoustical design. The design had to allow for a system of variable acoustics to accommodate the cinema situation.

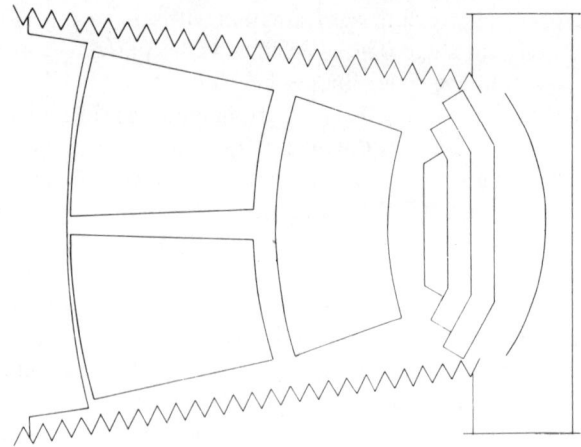

FIG. 3.11. Reykjavik University Hall (Háskolabio)—plan.

The main shape was a very modest fan shape and an essentially uniform ceiling height, with the exception of the stage tower. The architect had strong preference for a peculiar shape of the boundaries. He wanted to use a strong saw tooth profile in concrete for both walls and roof, thus shaping the structural surfaces to carry load as well as to provide stiffness. As a matter of principle, he also wanted to see this structure exposed on the inside of the auditorium. This indicated that highly diffusing surfaces would dominate both side walls and ceiling.

It was easy to accept this design acoustically, once the idea of using the ceiling as 'a sound distributing device' had been discarded. With regard to the walls it was suggested that use should be made of the saw tooth shape in

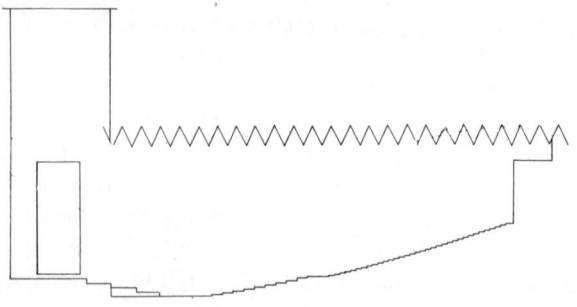

FIG. 3.12. Reykjavik University Hall—longitudinal section.

FIG. 3.13. Reykjavik University Hall—interior view towards stage.

FIG. 3.14. Reykjavik University Hall—exterior view.

connection with a system of variable acoustics. It was possible to apply panels hinged vertically along one edge and to give the two sides of the panels different acoustic treatment—one side with perforated panels covering absorbent material, and the other side with unperforated panels. The two different positions in which the panels were flush to a wall section would then represent different acoustical conditions.

It might be expected, however, that at low frequencies the difference in absorption for the two positions would be rather low. The variation in RT to be expected, at medium and high frequencies, in the occupied auditorium was from 1·1 to 1·6 s and the measurements in the completed auditorium showed that this variation was obtained.

The stage tower had to be shielded from the stage when the hall was to be used for concerts, and the same applied to the side wings of the stage. The fly tower was shielded off by a system of suspended horizontal plexiglass reflectors, which could be moved to a vertical position when not in use.

The side wings were shielded by a decorative curtain of heavy plastic material (lead vinyl) but it was agreed that this curtain should be supplemented with wooden barriers at the sides.

During the opening season some of the orchestra members voiced complaints concerning the hearing conditions on the stage. The stage was actually much larger than required and the orchestra arrangement was not quite satisfactory. After a few seasons it was decided to complete and improve the stage arrangement.

First of all, the wooden barriers were established and next, an orchestra shell was installed to provide more sound reflections for the musicians. The lead vinyl curtain behind the stage was rearranged and made smooth in appearance, instead of wavy, to avoid absorption of high frequencies.

The checking of acoustical changes was carried out by testing Early Decay Time (EDT) before and after the mentioned alterations. EDT is a criterion for acoustical quality which was developed some years later than the criterion of rise time. The story of this development will be told in a subsequent chapter (Chapter 4). Values of EDT may be evaluated by comparing them with values of RT and observing any major variations. Values of EDT which are considerably lower than values of RT are regarded, generally, as signifying acoustical deficiencies.

The testing of EDT before and after alterations indicated a change of the relation between values of EDT and values of RT over a broad range of frequencies (Fig. 3.15). The subjective reactions of the members of the orchestra also seemed to confirm that the acoustical properties of the stage

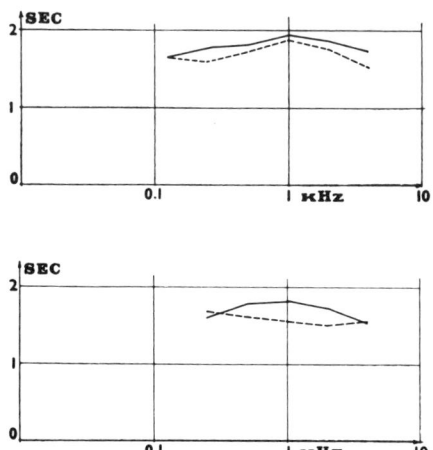

FIG. 3.15. Reykjavik University Hall—RT (s) (top) and EDT (s) (bottom) v. frequency (kHz) curves. —— enclosed stage, --- open stage.

had improved. Principally, the alterations must have had the effect of increasing the amount and the diffusivity of the early reflections in the stage area.

In the case of the Reykjavik Hall the combination of the two (opposed) purposes of concert hall and cinema has given a fair result. It is obvious that the much more spacious layout, compared with the Scala Theatre, has contributed to this. With the gentle slope of the Reykjavik Hall and the considerable ceiling height it has become possible to create and sustain a reverberant field in the horizontal direction, in the concert hall situation. The mechanical system for moving the hinged side wall panels has worked reliably and it is easy to handle. On the rear wall this system is supplemented with movable curtain strips, not unlike the system used in Aarhus.

Working with concepts like rise time (TR) and early decay time (EDT) marks some of the stages in a long development of acoustical criteria. It is a development which, in its origin, is closely associated with experiences of shortcomings in completed auditoria, but which, in its later stages, becomes more deliberately directed towards a definite goal: to establish acoustical criteria correlated with subjective assessment of acoustical quality. Such criteria should not be too complicated in terms of testing procedure and measuring equipment, and most important, should be those which could also be measured in scaled models of halls, using the same procedure, before the halls or theatres are actually built.

BIBLIOGRAPHY

Subject	Reference
Aalborghallen—general description	Jørgensen, J. *Ingeniøren*, **62**, 1953, p. 361.
Aalborghallen—acoustics	Jordan, V. L. *Ingeniøren*, **62**, 1953, p. 361.
Scala Theatre, Aarhus	Møller, C. F. *Arkitektur*, no. 2, SAR Förlag, Stockholm, 1975.
Reykjavik University Hall	Jordan, V. L. *J.A.S.A*, **47**, 1970, pp. 408–412.

DATA FOR AALBORGHALLEN, AALBORG

Year of completion: 1952/53

	Volume (m³)	Area (m²)	Number of seats
Assembly hall:	30 000	2 840	3 200
Concert hall:	25 200	1 800	1 800
Opera/Drama theatre:	18 200	1 310	1 400

RT measured in concert hall with capacity audience (octave filter):

63	125	250	500	1 000	2 000	4 000	8 000	(Hz)
2·2	1·7	1·7	1·8	2·0	2·1	2·0	1·4	(s)

General Description

side walls: (perforated) wood panels (10 mm), air space; mineral wool

rear walls: (perforated) wood panels (10 mm), air space; mineral wool; upper part of rear walls, angled reflectors

ceiling: wood panels (20 mm)

floors: parquet

platform walls: wood panels (20 mm), hinged

platform ceiling: wood panels (20 mm), hinged

seats: upholstered seat and back

platform size: 102 m²

DATA FOR SCALA THEATRE, AARHUS

Year of completion: 1953
Volume: 5000 m³
Total area: 600 m²
Number of seats: 800

RT (medium and high frequency range) measured with capacity audience:
Concert hall: 1·5 s
Cinema: 1·1 s

General Description

side walls:	concrete (covered with curtains when used as cinema) front of wooden grill
rear walls:	concrete
ceiling:	wood panelling
floors:	parquet
stage ceiling:	wood panelling (with slots)
seats:	upholstered
platform size:	95 m²

DATA FOR UNIVERSITY HALL, REYKJAVIK

Year of completion: 1962
Volume: 9700 m³
Total area: 880 m²
Number of seats: 1000

Concert Hall situation
RT and EDT, measured without audience (after adjustments):

	125	250	500	1 000	2 000	4 000	(Hz)
RT	1·65	1·77	1·81	1·93	1·86	1·74	(s)
EDT	—	1·60	1·78	1·82	1·72	1·54	(s)

General Description:

side walls:	hinged panels, perforated plywood on one side, unperforated on the other side, 25 mm mineral wool filling
rear walls:	wooden grille, air space, concrete

ceiling: concrete
floors: rubber tiles
stage ceiling: plexiglass reflectors
stage walls: (thick) wood panels
seats: upholstered
platform size: $180\,\text{m}^2$

Chapter 4

Development of Criteria and Model Research, I

Considering the acoustical design of concert halls and theatres the present situation cannot be properly understood and evaluated without taking a look back towards the origins of the development in this field. As an applied science, room acoustics is less than one century old. The art of theatre design has an interesting history which goes back to the Greek amphitheatres and can be seen as a continuous development over the Medieval and Baroque types of theatre as reviewed in Chapter 2. But this history is only very loosely associated with acoustical design considerations. Also, the classical concert hall design developed without the assistance of such considerations.

Examples of deviations from the classical rectangular shape of a concert hall were scarce and a case like the Royal Albert Hall has very little to do with rational application of a science; almost the contrary.

The fundamental concept of room resonance was understood and explained by Lord Rayleigh in his publications less than a century ago, and from his investigations it was evident that pronounced resonances appear almost exclusively in small rooms whereas in large concert halls the natural modes are so abundant in number that they lose individuality.

The real breakthrough of room acoustics as an applied science came with the energy concept (of W. C. Sabine) which explained the transient behaviour of acoustical energy within an enclosure and the approximate exponential decay of a sound field. By his theoretical investigations and through his practical work as an acoustic consultant Sabine created a tradition which lasted almost two generations and which concentrated practical room acoustics on the calculation of reverberation time and also, as the measuring technique improved, on the measurement of this parameter.

With the increasing possibility of predicting the value of RT on the basis of architectural drawings of a hall and on the basis of an increasing knowledge of the absorbing properties of materials, the interest soon concentrated on the problem of what numerical value should be chosen for RT in halls of different sizes and for different purposes.

For a concert hall various suggestions were proposed but a consensus did not exist, whereas for assembly halls for speech it was easier to establish definite values of RT especially as experience of intelligibility testing became available. The Sabine tradition of calculation and measurement of RT meant that the problem of shape, and geometric concepts of sound rays, did not receive so much attention. But architects, who are always concerned with the visual aspect and who are also more familiar with geometric concepts than with calculations, revived the interest in this aspect of shape by proposing, and in fact building, new halls of new shapes and abandoning the simple rectangular shape.

The first example, Salle Pleyel in Paris has already been mentioned (Chapter 1). The aim of this design was not to provide the correct reverberation but strength and proper distribution of sound to a large audience. While acousticians worried about the proper amount of reverberation, architects favoured solutions which concentrated on quite different aspects. The resulting confusion left its mark on a generation of halls completed during the thirties and forties. Gradually, it became increasingly evident that RT was not the universal criterion which the Sabine tradition imagined and that in large halls the listener evaluates the acoustics according to other criteria.

Thorough and systematic research after the war indicated that the analysis of the transient phenomena associated with very short sound pulses radiated in a large hall gave important clues to the problem of what the human hearing distinguishes as acoustically relevant. The reverberation process, i.e. what happens *after* a short transition interval, has not lost all its importance but the (Sabine) RT became one factor among several others.

When describing the development of criteria the emphasis will be on those with which the author has a personal experience, and also on those associated with music rather than on speech, without neglecting the latter completely.

An early criterion for speech was proposed, by R. Thiele in 1953, and termed 'Deutlichkeit' (definition). This is defined as the ratio of 'useful' energy (from a short pulse) arriving within the first 50 milliseconds to the total energy. The choice of the interval, 50 milliseconds, was made because this would be a reasonable figure for speech articulation. At that time very little was known concerning which interval would be appropriate with regard to musical sounds. This led many to use the same criterion for music, although it was proposed explicitly as a criterion for speech.

An early criterion for music has already been mentioned in connection with the measurements undertaken in the concert studio (of Radiohuset, Copenhagen, see Appendix I). This is the rise time (TR, the pulse length required to reach an energy level 3 dB below the stationary level). The elaborate method applied for measuring TR was an incentive to look for other criteria associated with the phenomenon of energy build-up in a hall. Imagine that this build-up curve (energy v. time) could be characterised by the steepness of the curve at a certain energy level. If the sound signal is a noise pulse of considerable length (e.g. 0·5 s) and if a high-speed level recorder is used to record the build-up curve (10 or 25 dB range) then the steepness of the curve at a definite level would be a measure of the rapidity of the build-up process.

Curves of this kind were recorded in the concert studio and the top 10 dB part of this curve indicated an almost constant slope, disregarding the fluctuations associated with a random noise signal. These fluctuations depend upon the shape of the noise signal at the starting point of the pulse and to establish a reliable average value of the slope the same process has to be repeated several times. Figure 4.1 shows examples of build-up curves. The method, therefore, is hardly less elaborate than the method used to measure rise time.

When using an interval of 10 dB of the build-up curve the slope was measured at the point 5 dB below the stationary level, which was a convenient choice and which was also considered relevant for the purpose.

The slope of the curve was defined as *Steepness* (σ) and was measured in dB/ms. To check the experimental values, they were compared with a theoretical value, calculated under the assumption of a perfectly logarithmic decay and absolute complementarity between build-up and decay process. This resulted in a simple formula (see Appendix II) according to which the theoretical steepness is expressed as:

$$\sigma \sim 0.13/RT \quad \text{(dB/ms, when RT is measured in seconds)}$$

The measured values of steepness could be expressed as a percentage of this theoretical value. This was carried out for the measurements made in the concert studio and the results were also compared with the values of rise

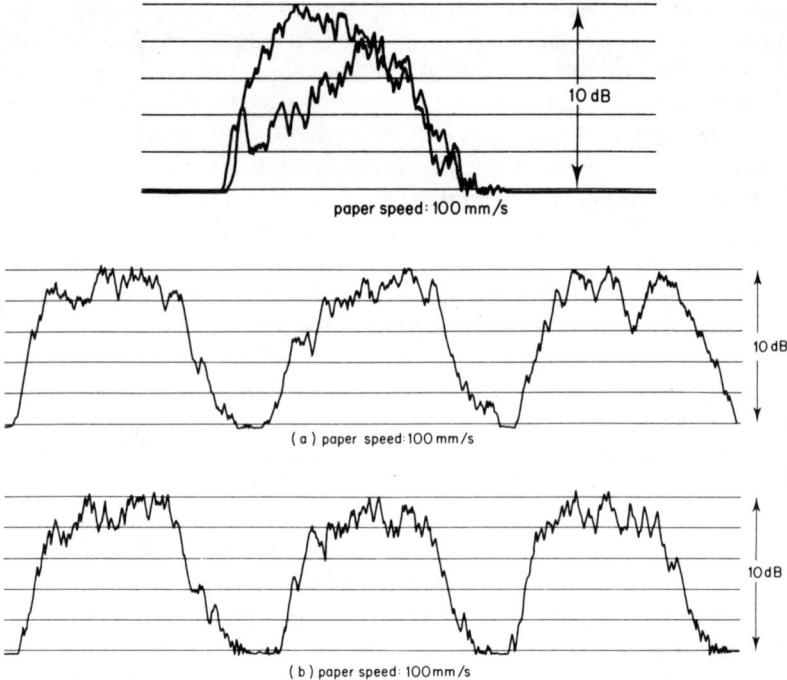

Fig. 4.1.   Concert Studio (Copenhagen) and New York State Theatre—examples of build-up curves. Top—Concert studio. Pulse length: 250 ms. *Upper track:* location at the orchestra level; *lower track:* location at the stage. Reverberation time ∼1·5 s. Bottom—State Theatre. Pulse length 50 ms. (a) In 1/10 scale model of the theatre. (b) In the auditorium of the theatre. For (a) and (b) the signal pulses used are identical; three successive pulses are shown. Location is orchestra, rear, in both cases. Reverberation time ∼0·16 s in model and ∼1·6 s in the hall.

time, also expressed as a percentage of the theoretical value. Table 1, in Appendix II, indicates some of the results.

To express the fundamental idea which has been mentioned earlier, that the build-up of the energy on the stage of a concert hall shall be for preference more rapid than in the audience area, the concept of *inversion* and the definition of an *Inversion Index* (II) may be introduced. In the case where we are using the criterion rise time, the inversion index is defined as:

$$\mathrm{II} \sim \frac{\text{average value of rise time in audience area}}{\text{average value of rise time on the stage}}$$

In the case where we are using the criterion steepness, the inversion index is defined as:

$$II \sim \frac{\text{average value of steepness on the stage}}{\text{average value of steepness in audience area}}$$

Values of II calculated from the measurements in the concert studio were in both cases considerably below 1·0 (for rise time 0·63 and for steepness 0·57, see Appendix II). The figures illustrate the fact that steepness may be more critical than rise time when checking on inversion phenomena. Later, when we introduce the criterion early decay time (EDT) we shall see that this criterion is *less* critical than rise time, when checking on inversion.

Before going further into the development of criteria, let us consider the application of scaled acoustic models and how this application has developed. We shall only consider physical (geometrical) models filled with atmospheric air. The interest for models of this kind goes back to the thirties although the limitations of measuring equipment considerably restricted the application of models.

Theoretical considerations showed that due to the sound absorption in air at high frequencies the similitude between a model and the corresponding hall would be severely limited in frequency range if a scaling factor of more than 1/10 or 1/20 was used. Increasing the factor to 1/5 or 1/8 would be preferable in this respect but the size and cost of models would also increase prohibitively, and so would the space requirements for housing such models.

One of the crucial questions is whether one wishes to use a drying-out procedure of the air in the model. By drying out the air down to an extremely low relative humidity (2–3 %) the sound absorption of the air at high frequencies is reduced so much that correct values of RT can be achieved in the model over the major part of the frequency range. This makes it feasible to reproduce samples of musical programmes in the model and listen to the results. Assessing acoustical quality from such musical samples requires elaborate investigations of groups of listeners and is not suitable for practical testing. It seems more expedient to limit model testing to the measurement of objective criteria. It is also extremely impractical to work with drying-out procedures which must be repeated each time it is necessary to test any changes made in the model (even when changing positions of sound source or microphone).

Testing objective criteria also makes the limiting of the frequency range less critical because it is possible to obtain valid results by measuring within octave bands, for example. In the early stages of model research some research workers confined their investigations to the observation of oscillogrammes of short pulses without being too much concerned about whether the model boundaries had approximately the correct absorption.

It was considered wise to go one step further: to aim at a model approximation which, within the octave bands tested, had a fairly accurate absorption (i.e., fairly accurate value of RT), but without paying too much attention to the upper range of frequencies, where the absorption was dominated by the absorption of the air. The more ambitious scheme of reproducing musical programmes over two-channel listening systems was left out of consideration.

The decision favoured a model scale of 1/10. Models of this scale are still expensive to build but when the drying-out process is omitted they are easy to handle with regard to changes of shape and details, which generally in practice will be the aim of the study. A question which had to be considered was the problem of absorption in the model, but since normal conditions in a concert hall mostly imply reflecting surfaces (with one major exception— the audience) the problem could be reduced by applying reflecting materials to the model surfaces (at the model frequencies). Absorption by (hard) panelling and of simulated audience could conveniently be tested in a model reverberation chamber. It was found, for example, that hard fibre board (with two coats of varnish) would have an absorption coefficient in the model (medium) frequency range similar to that of wood panelling, or other hard panelling, in the normal (medium) frequency range. Various types of cut neoprene foam would have absorption figures at model frequencies similar to those of a seated audience at normal frequencies. By choosing specially tested designs of model auditors—wood covered with thin cloth—even better approximations could be achieved. It became standard procedure to use neoprene blocks with cardboard heads, to simulate the diffusion around the heads of auditors.

The basis for instrumentation became a specially designed high quality tape recorder with two normal tape speeds in the exact 1/10 relation. As a sound source for the model testing, various types of loudspeaker have been used, one of them consisting of 12 small electrostatic speakers mounted on a sphere, and fed with pulses of random noise.

Later, a specially developed high voltage spark gap was applied to produce short sound pulses; this device became especially useful after the introduction of the method of integrated impulse measurements. This method, incidentally, reduced the variations of build-up curves (as well as decay curves) so that, e.g., measurements of steepness required only a few repeats.

The development of the model technique as outlined goes back almost 20 years, and the first major project where model testing was considered was the early designs of the two large halls of the Sydney Opera House. The largest of the halls, at that time called 'Major Hall' was, according to the programme, classified as an opera/concert hall, seating 1800 for opera and 2800 for concerts.

The original idea of the architect (Jørn Utzon) was to create one large enclosure, under the exterior shells, which could be either an opera theatre with the audience in front of a proscenium (and with a full stage and fly tower behind the proscenium), or a concert hall, where the audience seating was extended to occupy also the whole of the stage floor and where an immense, mechanically operated, ceiling screen would close off the fly tower. This was a great idea which, however, never materialised, but on which the early designs of the major hall were based.

The very first model of the major hall, built in hard fibre board sheets on a wooden framework, had the 1/10 dimensions of the intended concert hall interior. The audience simulation was primitive—fibre-glass blankets, 10 cm thick, covering all audience areas. This model could only be accommodated when one side wall rested on the floor of the laboratory and floor and ceiling spans of the model were orientated vertically. The objective testing method was not fully developed at that time; the first attempts were to apply the method of measuring rise time (Chapter 1). Directional effects of electrostatic membrane speakers tended not to give reproducible results and this led to the introduction of dodecahedron speakers. A complete change in the architect's intentions, concerning the design of the major hall, made this model obsolete.

Meanwhile an important project for a large ballet theatre had turned up: The New York State Theatre (of Lincoln Centre; architect, Philip Johnson). This theatre, with a seating capacity close to 2800, would be an object well suited for model testing. The testing technique had now developed and measurements of steepness in the Concert Studio of Copenhagen had shown possibilities of practical application.

The principal design of the auditorium (Figs. 4.2, 4.3, 4.4 and 4.5) could be considered almost finalised at the time when the model was being built, but some questions, e.g., ceiling shape and details of wall shape, were unresolved and could be studied in the model. The wall shape of the auditorium needed particular attention since the concave shape of side walls and rear wall could create an uneven sound distribution over the audience seating at floor level.

It was suggested that a shape with stepped panelling should be

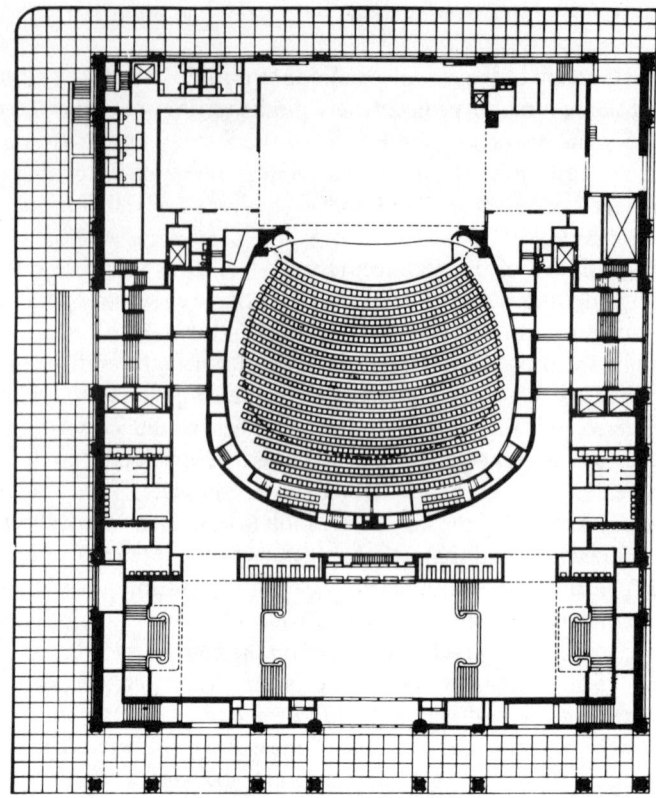

FIG. 4.2.   New York State Theatre—plan of the orchestra level. The panelling of
the side walls is not shown.

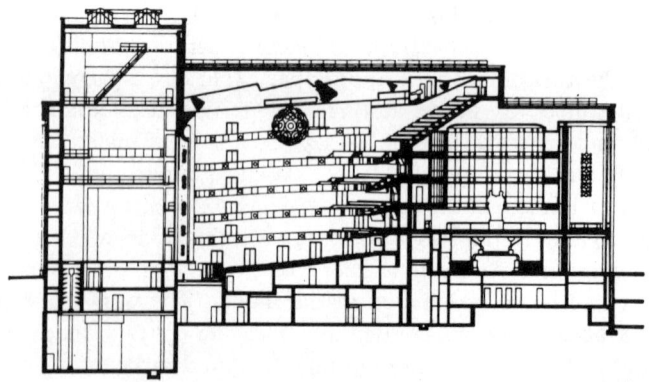

FIG. 4.3.   New York State Theatre—longitudinal section. The reflecting ceiling
(above the grill ceiling) is indicated, as are the loudspeaker locations.

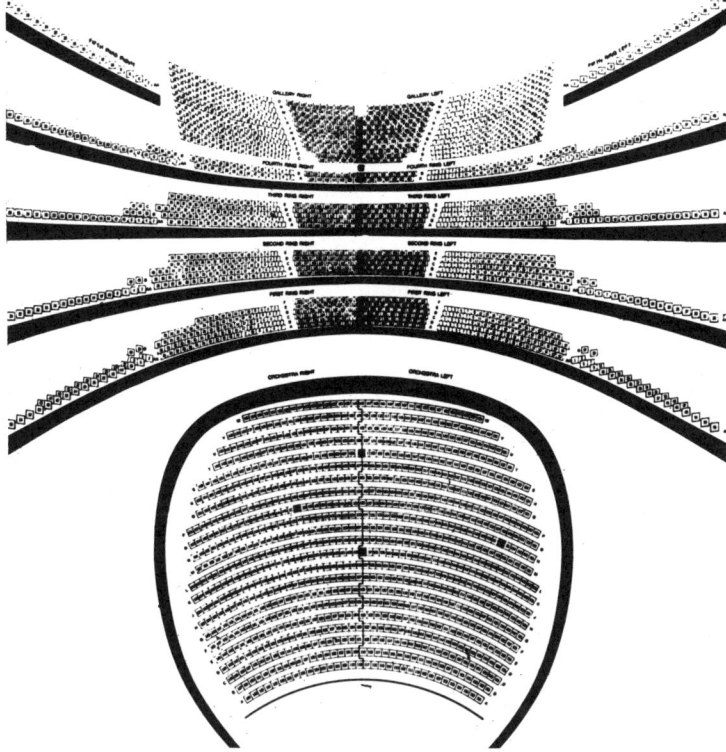

FIG. 4.4.   New York State Theatre—layout with microphone locations indicated by black squares.

FIG. 4.5.   New York State Theatre—view of the interior of 1/10 model.

introduced which would diminish the focusing tendencies of the walls, for medium and higher frequencies.

Measurements of steepness in the model of the auditorium indicated quite a large variation of numerical values with the position of the microphone and the values also indicated a certain correlation with the specific individual locations. In addition, the values were comparable with the calculated value (calculated from measured values of RT).

The model was measured on several occasions (Fig. 4.6) and two typical situations should be considered here. At an early stage the RT value was 0·19 s (corresponding to 1·9 s in the future hall); with more details included in the model the RT value was 0·16 s (very close to the expected value in the hall, 1·6 s).

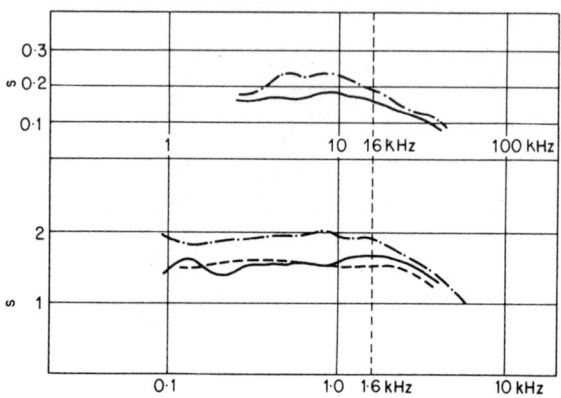

FIG. 4.6.   New York State Theatre—RT v. frequency curves for model (top) and auditorium (bottom). —·— early stages; —— final stages; — — — auditorium with 90% audience.

The steepness values in a number of locations are given in Table 2 of Appendix II which also indicates the calculated values. Although random variations still exist, the relative evaluation of the different locations is quite well expressed by the values of steepness. The lowest values are always measured at the centre of the orchestra level. This may be taken as a warning about the problems which may actually appear in the theatre auditorium at corresponding locations.

At an early stage of the project advice had been given that a large overhead reflector should be installed in front of the proscenium wall but the idea was discarded for architectural reasons.

Primarily, the theatre was to be used for ballet, but as a secondary function it would be used for musicals, operetta and light opera. Particularly in the case of musicals the decision was taken to rely on artificial support for the voices. The acoustical testing, however, went on, first in the model as described above and later in the hall itself. Prior to the inauguration in 1964 the auditorium was tested on two occasions (again with slightly different values of RT) and the testing equipment was in principle the same as used in the model testing. It is interesting to compare the values measured for the steepness (see Appendix II).

Even if individual results do not coincide when comparing model and auditorium, it is evident that extreme positions (like centre positions and wall positions) are also characterised by extreme values of steepness and that this occurs in both cases. It also appears that the final series, in both cases, are more consistent when compared than are the first series (maybe because of the more diffuse sound fields in the final series). Average values of steepness are also closer to 100 % of the calculated value in the final series.

The criterion of steepness was also applied in other cases of model investigations. Looking at the formula for steepness:

$$\sigma = 0{\cdot}0094 c \alpha_m / p \quad (\text{dB/ms})$$

where c is the velocity of sound, α_m the mean absorption coefficient and p the mean free path, the expression indicates that the value of steepness would be expected to increase with a decrease in p. When one considers that the sound paths which are effective in determining the steepness are those which occur early in the build-up process and include the reflections with short time delays, it would seem possible to influence the value of steepness. If, for example, reflecting surfaces were placed in some of the sound paths between source and receiver (surfaces which were parallel or almost parallel) then this would correspond to a reduction of the effective mean free path in an early interval of the build-up process. In a specific case, this hypothesis was actually investigated in a model.

The model was a later version of the major hall of the Sydney Opera House. At this stage the original idea of locating a large part of the audience on the stage floor had been abandoned by the architect. The auditorium had been increased in width (but had nevertheless decreased in capacity to 2200 seats) and a permanent proscenium divided the stage from the audience area (Figs. 4.7, 4.8, 4.9 and 4.10).

It was suspected that large areas of the audience seating would have low values of steepness, due to absence of reflected sound, and it was decided to

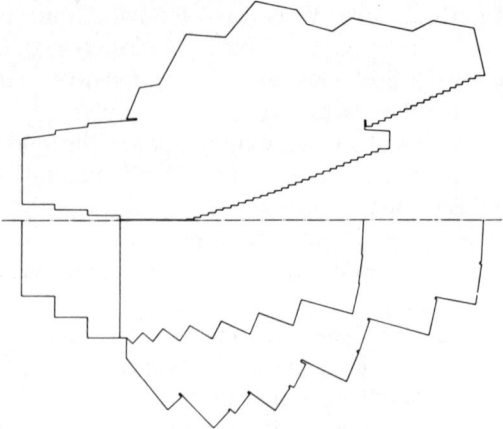

FIG. 4.7.   Major Hall, Sydney—last Utzon design, half-plan, and section.

test this by a model investigation. The first audience simulation was again the same primitive one as used in the model of New York State Theatre, but later the simulation was improved by use of individual neoprene foam cutouts.

These two cases were tested at eight different microphone locations and the comparatively low values of steepness at centre locations which were measured in the first case improved somewhat in the second, maybe

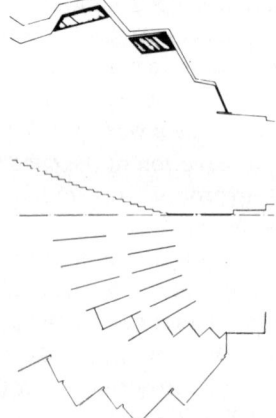

FIG. 4.8.   Major Hall, Sydney—last Utzon design. Section with vertical baffles indicated.

FIG. 4.9.   Major Hall, Sydney, model—view towards stage.

through added reflections from the rows. To test the idea of improving the steepness by reducing the mean free path, individual vertical reflectors were suspended in two rows under the ceiling (see Figs. 4.8 and 4.9).The result was a definite increase in the steepness values with the average now approaching the theoretical value. Table 3 in Appendix II states the measured values expressed as a percentage of the theoretical values.

FIG. 4.10.   Major Hall, Sydney, model—view towards audience.

The practical measurement of steepness was cumbersome and elaborate, especially before the introduction of the integrated impulse method. Furthermore, the importance of steepness for the subjective evaluation of acoustical quality was lacking in direct evidence although it was natural to assume that relatively large values of steepness should be favourable.

At that time certain suggestions of correlation between subjective impressions and objective criteria appeared in a now famous publication, by M. R. Schroeder and his co-workers. The correlation was indicated between subjective evaluation of reverberation and the objective value of 'Initial Reverberation Time', which was measured either over the first 15 dB of the decay process or over a fixed interval of 160 ms (the latter would come close to 5 dB if RT was close to 2 s). Taking this investigation into consideration it was decided to include a modified version of this criterion in future testing and to define this as Early Decay Time (EDT) which means the value of reverberation time corresponding to the slope of the first 10 dB of the decay process. The question, of course, arises: If the decay process (in logarithmic scale) is not linear, how do we measure the value of EDT? It has been a long established practice not to take irregularities in the first 10 dB part of the curve into consideration, but to measure the slope from the zero point to the −10 dB point, without worrying too much about theoretical implications. The justification for this is that the ear may not be particularly sensitive to irregularities within this interval.

The practical possibility of measuring this first slope of a reverberation process is intimately associated with the integrated impulse method, as theoretically and experimentally introduced by M. R. Schroeder. His investigations showed that the integrated impulse response represented the summation of an infinity of random decay processes, which with one stroke eliminated the laborious task of averaging over a large number of individual recording tracks of build-up or decay curves.

In another contemporary publication Schroeder showed that with certain assumptions, perfect reciprocity between build-up and decay curves in a room was to be expected. He also showed that large values of steepness measured on the integrated build-up curve indicated a surplus of reflections in a time interval of around 50 ms delay. This is a first indication linking steepness to the so called 'energy criteria' which was proposed by others (see Chapter 9).

Attempts to apply objectively measurable criteria to model testing (and to testing in full scale) continued in an investigation of a 1/10 scale model of the project for a new Metropolitan Opera of New York, situated at Lincoln

Centre. During this investigation, both new criteria, steepness and EDT, were measured and the testing equipment was developed. The technique of repeating random noise pulses to measure steepness was superseded by the integrated impulse response method, using an electric spark source and an integrating device. Both steepness and EDT could thus be measured; steepness by using forward integration of the impulse response and EDT by reversing the tape and using backward integration. The concept of inversion index (II), originally introduced in connection with rise time measurements as a sort of indication of balance between an orchestra platform and an audience area, was applied to indicate the balance between individual audience areas. The reasoning corresponded to the original idea that, ideally, audience areas close to the stage should have larger values of steepness than audience areas further away from the stage.

In the course of the investigation of the model of the new Metropolitan Opera the emphasis shifted from steepness to EDT and values of EDT were also compared to values of RT (for details of decay process, EDT and II, see

Fig. 4.11.   New Metropolitan Opera, model—view towards stage.

Appendix III; for model of the new Metropolitan Opera see Figs. 4.11 and 4.12).

The ambiguous situation with regard to the two main criteria applied (steepness and EDT) also arose in the early stages of investigations in another model, the model of the Oslo Concert Hall, which started ten years before the inauguration of the hall (in 1977). The investigations in this model were numerous but one of the first undertaken involved the question of the ceiling shape. The hall itself had a general shape approximating to a triangle, dictated by the shape of the site. This shape was much disputed and

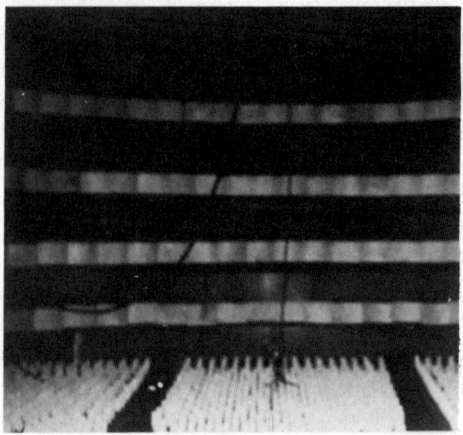

FIG. 4.12.   New Metropolitan Opera, model—view towards audience.

it presented a number of problems but they did not interfere with the early investigation of the ceiling shape.

Due to the building regulations the height of the hall was limited and the idea of including the required height of the roof structure in the interior volume of the hall had interesting economic possibilities. The structure which was suggested consisted of concrete beams spanning the ceiling but the programme for testing the ceiling shape envisaged three possibilities

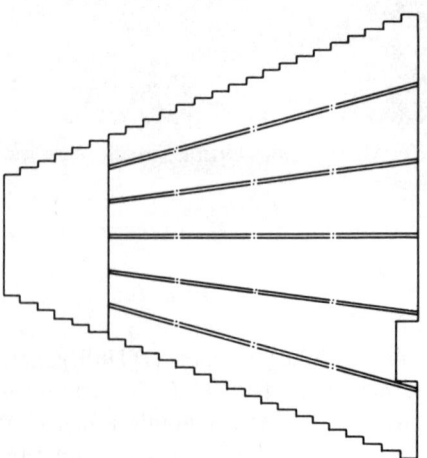

FIG. 4.13.   Oslo Concert Hall, preliminary design—plan.

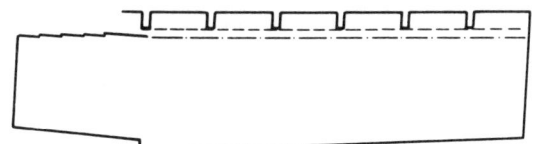

FIG. 4.14.   Oslo Concert Hall, preliminary design—longitudinal section.

(see Figs. 4.13 and 4.14): 1, a flat ceiling; 2, transverse beams and 3, radial beams (disregarding that 3, structurally, was out of the question). The testing method applied was now exclusively the integrated impulse method and both steepness and EDT were measured at a number of microphone locations. The important single parameter to observe was thought to be the inversion index as calculated from steepness and EDT values.

According to the usual evaluation of inversion index all values which are equal to or larger than 1·0 are considered favourable whereas values below 1·0 are regarded as indicating a basic imbalance in a hall. The calculated values of II from the measurements of steepness and EDT are given in Table 6 of Appendix III. Although values of II are not exactly the same, when calculated from steepness or from EDT, they are all in agreement concerning the evaluation of the different ceiling shapes. The shape with beams across definitely comes out as the best of the three different ceiling shapes.

At this point there was evidently a great need for resolving the problem of correlation between the applied objective criteria—EDT, steepness and II —and subjective assessments of musical quality. No doubt influenced by the previously mentioned publication on initial reverberation time and subjective assessment of reverberation, it was decided to give preference to the criteria EDT and II, and to apply these two criteria in the actual cases of model studies: first of all the Oslo Concert Hall project but also another major project, with two large halls, where the finalisation of design required extensive model investigations, the Sydney Opera House. This project deserves a chapter of its own so that the story may be traced back to the origin of the project (Chapter 6).

BIBLIOGRAPHY

Subject   *Reference*
Room resonance   Rayleigh, Lord John. *Theory of Sound*,
                            vol. II, Chapter XI. Macmillan and
                            Co., London, 1937.

Reverberation	Sabine, W. C. Collected Papers, Dover Publications Inc., New York, 1937, Chapter I.
Deutlichkeit	Thiele, R. Acustica, 3, 1953, p. 291.
Rise Time	Jordan, V. L. ICA III, Elsevier Publishing Co., Amsterdam, 1959, p. 922.
Steepness and inversion index	Jordan, V. L. Applied Acoustics, 1, 1968, p. 29.
Absorption of model audience	Day, B. Applied Acoustics, 1, 1968, p. 121. Hegvold, L. W. Applied Acoustics, 4, 1971, p. 237.
New York State Theatre—model testing, auditorium testing	Jordan, V. L. J. Aud. Eng. Soc., 13, 1965, p. 98.
Major Hall, Sydney Opera House—model test	Jordan, V. L. Acustica, 16, 1965/66, p. 187.
Metropolitan Opera, New York	Jordan, V. L. J.A.S.A., 47, 1970, p. 408.
Correlation between subjective reverberation and objective criteria	Atal, B. S., Schroeder, M. R. and Sessler, G. M. ICA V, Proceedings Ib, Elsevier Publishing Co., Amsterdam, 1965, G.32.
Integrated impulse response, new method of measuring RT	Schroeder, M. R. J.A.S.A., 37, 1965, p. 409.
Complementarity of sound build-up and decay	Schroeder, M. R. J.A.S.A., 40, 1966, p. 549.
Early decay time (EDT)	Jordan, V. L. Applied Acoustics, 1, 1968, p. 29. Jordan, V. L. Rundfunktechn. Mitteilungen, 13, 1969, p. 202.
Oslo Concert Hall—preliminary design of ceilings	Jordan, V. L. Applied Acoustics, 1, 1968, p. 29.

Chapter 5

New York State Theatre and Metropolitan Opera House, and National Theatres of Latin America

New York State Theatre

The classical horseshoe theatre shape has often been proclaimed as having specific virtues acoustically; on the other hand, there are quite a few examples of classical theatre auditoria which have been criticised as having built-in acoustical defects. Again, the clear distinction between theatre for the spoken word (drama theatre) and theatre for singing and music (lyric theatre) has not always been taken seriously enough. The question of size also comes in to the discussion and tends to confuse the issue.

An example of a theatre auditorium where the classical horseshoe shape was adopted by the architect both for aesthetic as well as for what we presume were acoustic reasons is New York State Theatre in the Lincoln Centre. The prototypes, no doubt, were the 18th–19th century City Theatres, for example in Germany, but in the early stages of the project no serious consideration was given to the fact that the mere change in size has quite substantial consequences, acoustically.

It was helpful, though, that the scope of this particular theatre, at the outset at least, was definitely limited to 'The Ballet Theatre' of the Lincoln Centre, leaving other purposes such as concert, grand opera and drama to other units of the centre.

An immediate addition to this clearly defined scope was 'light opera' and 'musical'. However, with the prime purpose being 'ballet', the main problem, acoustically, could still be considered a problem of musical acoustics and only secondarily a problem of articulation. The expansion from a normal European theatre of, say, 1500 seats to a capacity of 2800 seats, however, contributed to the acoustical problems. Just consider the elementary question of the time delay of the first ceiling reflections which instead of the normal 50–70 ms become of the order of 100–120 ms (corresponding to a height of 20–22 m). This delay will affect articulation considerably, and the secondary purpose cannot be totally neglected. Something had to be done about the first reflection from overhead.

The obvious proposal then was to suggest that a large canopy be placed above the stage opening to provide reflections from overhead with a more reasonable time delay.

The question might be posed as to why the solution applied a few years later to the Metropolitan Opera, of creating a broad proscenium frame (to give side reflections) was not brought into the discussion in the case of the State Theatre. The answer must be that the design of the proscenium and the surroundings was more or less 'frozen' and major changes were hardly feasible.

The suggestion of a canopy over the stage opening was adopted and the detailed design started along these lines. The design had to be tested by a model investigation in a 1/10 scale model to assess the acoustical parameters applied and to test the possible modifications of wall shapes. The main results of this investigation have already been mentioned (Chapter 4) and the tables of steepness values measured in the model (see Appendix II), indicate the improvement from the early stage (RT ∼ 0·19 s) to the final stage (RT ∼ 0·16 s) in which the details of panelling were added.

The details of the side wall panelling are in accordance with the often applied principle of stepping (parallel to the long axis of the hall) which has been used in this case not just to orientate the side reflections but also to do away with the concave shape of the walls. The improvement of steepness values in the orchestra seating area of the model may be clearly seen when looking at the average value from the four microphone locations in that area.

The improvement of steepness values in the same area in the auditorium itself is even more marked. Even if the exact subjective importance of the steepness has never been established it is safe to assume that an increase in steepness (whereby numerical values approach the statistical value) is favourable; especially so for articulation.

In the centre of the orchestra seating area the model measurements showed the lowest values of steepness, a fact which is also confirmed by the measurements in the auditorium. This deficiency was left for compensation by a reinforcement system.

The reinforcement system was designed to give an even distribution over all seats of the auditorium. The main speaker arrangement was two large rows of column speakers on each side of the proscenium, split up into vertical units which fitted in between the balconies. These column speakers covered most of the seating areas except the large rear extension of the 4th balcony ('the gallery'). To cover that area it was necessary to supplement the main system of columns with a horn speaker mounted in a centre position above the decorative ceiling (for loudspeaker locations see Figs. 4.2 and 4.3). This horn speaker, which was much closer to the audience than the actors on stage, had to be delayed (in time) so that the audience would orientate themselves towards the singers, and not towards the horn speaker. The decorative grill ceiling is of expanded metal and conceals the speaker but allows the sound to pass freely through.

Reporting three opening events which took place during 1964 will illustrate the ambiguous comments on acoustics which New York State Theatre received.

The first inaugural event was, of course, a ballet performance, staging Serenade (Tchaikovsky, Balanchine), Agon (Stravinsky, Balanchine) plus two American ballet pieces, choreographed by Balanchine. The comments on acoustics by conductor, orchestra, music critics and audience were unanimously favourable. The preliminary testing indicated that RT was close to 1·6 s and the RT v. frequency curve was practically flat (see Fig. 4.6).

The next major event, one month later, was drama performances by a British Shakespearean company using a *thrust stage*. This had *not* been part of the project and had *not* been envisaged when the reinforcement system was designed. The comments were sharply adverse. Articulation was bad in many areas and the acoustics were severely criticised.

The third major event, a month later, was a musical ('The King and I') performed on the regular proscenium stage with the singers using amplification but at a comfortable low level of reinforcement. The comments were reassuring and there were very few complaints about the acoustics of the auditorium.

The story is instructive. By completely disregarding the conscientiously implemented acoustical design criteria: 1, ballet; 2, musical and light opera, and by introducing drama (which had been totally discarded in the brief), an almost disastrous situation arose, which the general public and the press could easily interpret as: 'Something is rotten with the acoustics in the State Theatre'.

The following winter season (1964/65) had a regular series of classical opera performances, using the normal proscenium staging (only occasionally with reinforcement) and the reactions were favourable with no adverse comments on acoustics.

Fortunately, the usage of this theatre over the years since the

FIG. 5.1.   New York State Theatre—view of the auditorium.

inauguration has been almost exclusively for ballet and light opera. The incident with Shakespearean drama may today be regarded as a warning against careless management and against the abuse of an otherwise very well functioning lyric theatre. A view of the auditorium is shown in Fig. 5.1. Quite apart from acoustical considerations, who would dream today of playing drama to an audience of 2800 people?

Metropolitan Opera House of New York
The old Metropolitan Opera House of New York, between 39th and 40th Street, had a rather mixed reputation with regard to acoustics. Some of the favourable verdicts may have been influenced by the prestige and grandeur of this house more than by its acoustical properties. One thing is known for certain: it was a house where only the really big voices were powerful enough to make an impression and to fill the auditorium. Even the singers with big voices complained, because in order to be heard they had to force their voices, even in piano passages.

The old Metropolitan Opera was sometimes compared with other large opera houses of the world and the comparison was very often not to the advantage of the Metropolitan. Inevitably, it would then be added that, of course, the immense size of the Metropolitan auditorium was a drawback. However, it is questionable whether size in itself is a drawback. To make just one comparison: one of the most favoured opera auditoria of the world is Teatro Colon (in Buenos Aires, 1908). It has a volume which is larger than the old Metropolitan and yet it is acclaimed to be a most gratifying auditorium, especially for singers, but also for concerts (see Chapter 2). It has two additional features which must be of importance: it has much more volume per seat than the old Metropolitan ($8.4\,\mathrm{m}^3$/seat as against

$5 \cdot 4 \, \text{m}^3$/seat) but even more importantly, it has a shape which in plan will favour early side reflections.

When the design criteria for the new Metropolitan, at the Lincoln Centre, were established, they were only specific on one point: the number of seats. It was requested that the new opera house should have more seats than the old and of course the seating should also be more comfortable. No one wanted to commit themselves about the acoustics, more than to make vague utterances like 'As good as in the old house', or something similar. Referring to the classical criterion RT, the old house had $1 \cdot 2$ s (at 500–1000 Hz), which at once indicates that it was not exactly a 'live' hall. But even more instructive was the information of the RT v. frequency dependence (which is illustrated in Fig. 5.2). The lack of reverberance at high frequencies is remarkable.

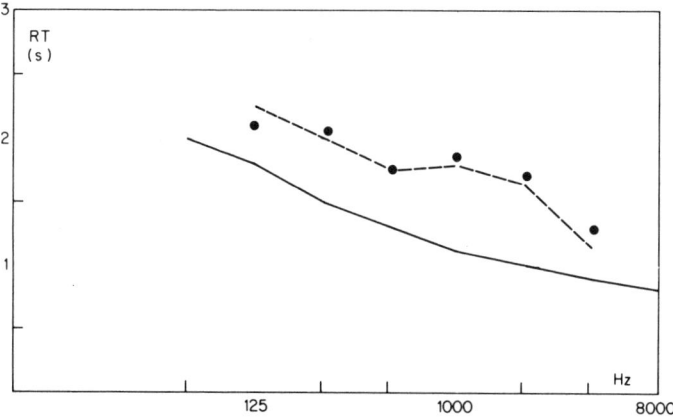

FIG. 5.2.   New and Old Metropolitan Opera, New York—RT (s) v. frequency (kHz): ———, old Metropolitan; – – –, new Metropolitan. EDT: ●, New Metropolitan.

When architectural design started (around 1959; architect, Wallace K. Harrison) several considerations on the general layout were followed by more specific acoustical suggestions. In the final period of acoustical advice, during 1961 and the following years, the author was mainly responsible for the adaptation of the architectural design to specific acoustical requirements (with due acknowledgement to his co-consultant, Cyril Harris).

The design had to be checked by model investigations in a 1/10 model (see Figs. 4.11 and 4.12). Of the several suggestions proposed (and accepted by the architect) the top priority was given to the proposal regarding how to shape the proscenium zone: to shape this zone as a transition space between the stage proper and the auditorium, to locate the orchestra at the bottom of this space and to design the vertical sides and the overhead arch as reflecting surfaces (Fig. 5.3), acoustically connecting stage area and audience area.

Early experiences (Chapter 1) may have assisted the author when insisting on advocating this design feature which may also be regarded as an

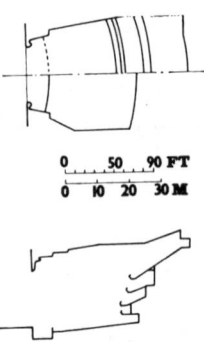

FIG. 5.3.   Metropolitan Opera, New York—schematic plan and section.

expedient to increase the diffusion for the singers as well as for the orchestra. However, without the acceptance and the cooperation of the architect these ideas would not have materialised.

The model testing concentrated on the measurement of two criteria, steepness and EDT (for definitions and formulae, see Appendices II and III). The evaluation of measured values followed along the lines already established, i.e., comparison of steepness values with calculated value and comparison of EDT values with measurements of RT. Both criteria were also applied to the calculation of inversion index. If applying the concept of II by evaluating the balance between different audience areas it appears that in this case a slight inversion exists between the orchestra seating area and the upper balconies (see Table 4 in Appendix II). This, indeed, may be difficult to avoid altogether in any auditorium of classical theatre shape. A check was made of whether EDT values at different locations were close to RT values (Table 5, Appendix III).

Later, the model measurements, both of steepness and EDT, were repeated in the finished auditorium so that comparisons could be made. Absolute values of EDT and RT were relatively lower in the model than in the finished hall but EDT values came close to RT values in both instances.

It should also be noted that percentage values of steepness are close to 100% both in the model and in the hall. Finally, when looking upon the measured values in the finished auditorium (occupied by 3000 school children at a test performance) it should be noted that the octave values of EDT and RT are close to each other over the frequency range 125–4000 Hz (Fig. 5.2). A direct comparison of the RT v. frequency curves between the old and the new Metropolitan Opera is especially interesting (Fig. 5.2). These two curves, so different in their frequency balance, explain a lot about the blatant difference in 'brilliance' (upper range reverberance) of the two auditoria.

Another point which concerns the benefit of the early side reflections in this auditorium is the fact that these reflections, arriving within a relatively short time delay, are useful, not solely for the musical quality but also for the articulation. That this auditorium has very good articulation for the spoken word was soon realised; in fact, at the test performance just mentioned. When the representative of the Metropolitan Opera Company was addressing the audience (with a clear and loud voice, but without any reinforcement), from his position at the front of the stage, it was immediately apparent that his speech was very well understood over the entire seating areas. Views of the auditorium are shown in Figs. 5.4 and 5.5.

The opening event, in 1966, was not particularly well suited for enabling the singers and the audience (nor the orchestra and the critics) to assess the acoustics of the hall. It was a modern piece by Samuel Barber ('Anthony and Cleopatra', composed for the inauguration) but even so there was a favourable indication when Leontyne Price, who had the leading part, expressed her warm appreciation of the acoustics that same evening after the performance. Another remark, worth retaining from the opening period when a series of classical operas was performed, is the relieved utterance by Renata Tebaldi, 'Now I can start singing softly again!'

During the opening season it became evident that this opera theatre really has remarkably good acoustics. It may be concluded that a number of essential design features contribute to this result (apart from the proscenium design). The downward reflecting surfaces of the balcony fascia, the thoroughly diffusing ceiling shapes, the meticulous avoidance of any surfaces absorbing the high frequencies; these features may all have influenced the undisputed good result.

Even if the insistence may appear too emphatic, it may not be possible to overestimate the influence of the basic configuration of the proscenium. As related above (and in Chapter 2) it is a feature belonging to the very best opera houses of the Baroque period, and later imitations.

FIG. 5.4. Metropolitan Opera, New York—view towards stage.
(Photograph courtesy of J. W. Molitor.)

FIG. 5.5. Metropolitan Opera, New York—view towards auditorium.

Looking back, it becomes evident that through the research on auditoria acoustics of the last 10–15 years everybody has become increasingly aware of the specific influence of early reflections (especially lateral reflections) on the acoustics of large halls, so that seen in retrospect the acoustical design of the new Metropolitan Opera corresponded well with growing tendencies of the art.

NATIONAL THEATRES OF LATIN AMERICA

The Ruben Dario Theatre of Nicaragua

Architects from Latin America were attracted to and impressed by the cultural centres of the sixties, of the USA and Canada. It is no exaggeration to say that one of the centres which drew most attention was the Lincoln Centre of New York. The centre now comprises five large units of which two have been previously described. (No comments are presented for the other three: the Avery Fisher Hall (formerly Philharmonic Hall), the Vivian Beaumont Theatre and the Juillard School of Music. The literature on Philharmonic Hall is abundant.)

The dazzling exterior and interior of the New York State Theatre inspired a young architect from Nicaragua, José Teran, and eventually this inspiration was the starting point for a project, the National Theatre of Nicaragua (named, after a national poet, the Ruben Dario Theatre). His consultants were part of the team from the New York State Theatre (consultant on theatre architecture, (the late) Ben Schlanger of New York). The capacity of the Ruben Dario Theatre was 1400 seats, far less than its model at the Lincoln Centre. Here the single unit would have to serve alternating purposes. The design criteria included stage performances of folklore, ballet and opera, as well as symphonic concerts—from orchestra to quartet. The chosen shape of the auditorium was favourable; close to rectangular, with seating at floor level and three balcony levels (Figs. 5.6 and 5.7). The shallow side balconies were terminated at some distance from the stage thus creating a transition zone which in this case was also designed as a forestage, with possibilities for flying stage sets. When using the hall as a concert hall a complete orchestra shell was to be wheeled forward on the stage so that reflecting surfaces were established on three sides of the orchestra.

The opening of the Ruben Dario Theatre in 1969 took place without any warning to the consultants. Shortly afterwards it was communicated to the author that the acoustics had been most favourably commented upon by several musical groups (visiting ensembles from Europe) and an opportunity was offered to check on the acoustical data of the auditorium.

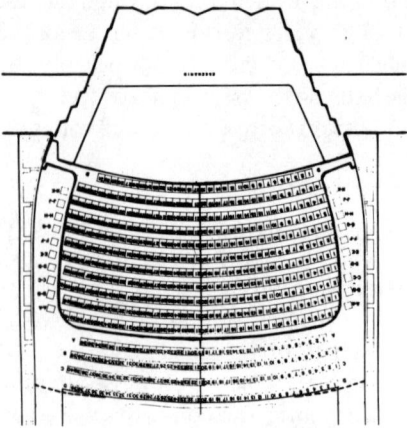

FIG. 5.6.   Ruben Dario National Theatre—plan.

It should be mentioned that the design of the ceiling includes some specific acoustical features. The suspended decorative ceiling, of expanded metal and similar to the ceiling of the State Theatre, conceals several rows of vertical baffles, parallel to the long axis of the hall. The idea was to increase diffusion of the ceiling reflections, more specifically to increase the amount of laterally distributed energy. This is the same idea as described for the Major Hall design (Sydney Opera House) which was tested in a 1/10 model and where the testing had shown that the arrangement of vertically suspended baffles may actually cause an increase in values of steepness (see Chapter 4 and Appendix II, Table 3). Another design feature of the

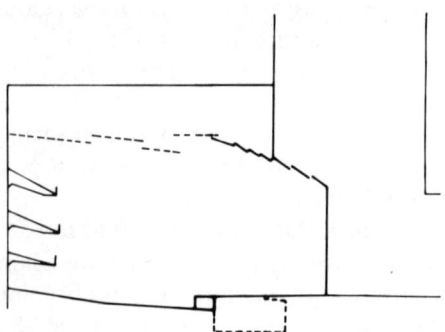

FIG. 5.7.   Ruben Dario National Theatre—longitudinal section.

auditorium is the convex, reflecting balcony fascia which together with the
side walls provides lateral reflections at floor level.
The testing of the acoustics of the auditorium was therefore anticipated
with great interest. The criteria tested were: RT v. frequency; EDT v.
frequency and EDT at different locations; and impulse response level at
different locations. RT and EDT v. frequency curves are shown in Fig. 5.8.

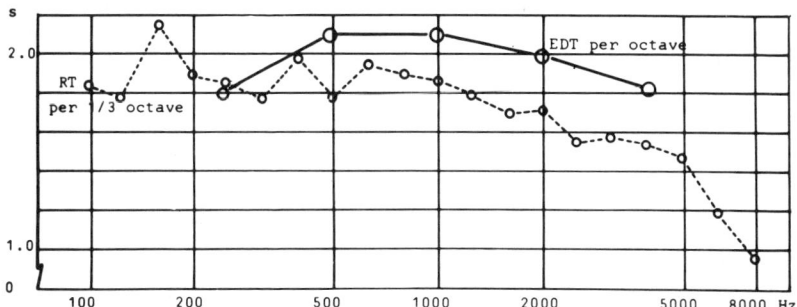

FIG. 5.8.   Ruben Dario National Theatre—RT and EDT v. frequency. -----, RT;
——, EDT.

Impulse response levels as well as EDT values are indicated in Appendix IV,
Table 10. From the frequency curves of RT and EDT it is noted that values
of EDT (per octave) are somewhat in excess of values of RT (per 1/3 octave).
Impulse tests do sometimes exhibit differences of this kind, i.e. variations
with the width of the band filter. At any rate, values of EDT are at the same
level as values of RT. The distribution of EDT values (at the octave of 2 kHz)
over the different locations shows remarkably small variation and the same
applies to the values of impulse response level (measured with a B & K
sound level meter, impulse response). The inversion index is practically 1·0.
All testing was carried out in the empty hall and with the orchestra shell in
position. The test results may be said to support the subjective assessments
of the various musical groups.

The National Theatre of Guatemala
A more recent theatre building in Latin America is the National Theatre of
Guatemala. The building itself, however, has a rather long history since the
original project goes back almost 20 years. The basement and the podium
(including the stage floor) was built and lay in bare concrete for almost a
decade before construction was resumed in 1972 (architect, Efrain
Recinos). The design criteria as agreed between the management and the

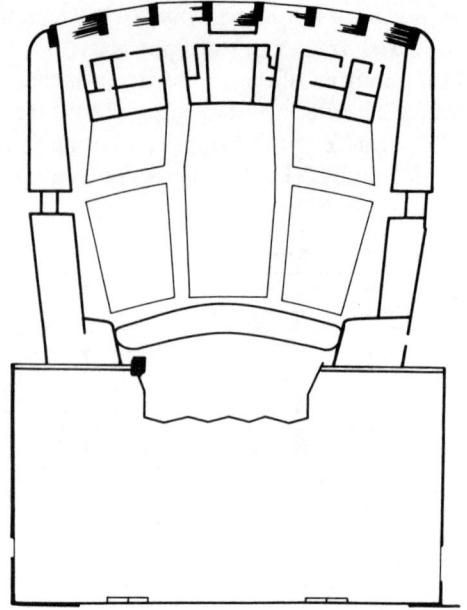

FIG. 5.9.   National Theatre of Guatemala—plan.

consultant at that time were: 1, concert hall; 2, opera (full stage and pit facilities); 3, folklore ballet; and 4, mass meetings. The layout in plan and section could still be modified in accordance with acoustical preferences and three major changes of the project were proposed and accepted: 1, the side walls were brought closer to each other; 2, the ceiling height was increased by eliminating a solid ceiling at the bottom level of the steel

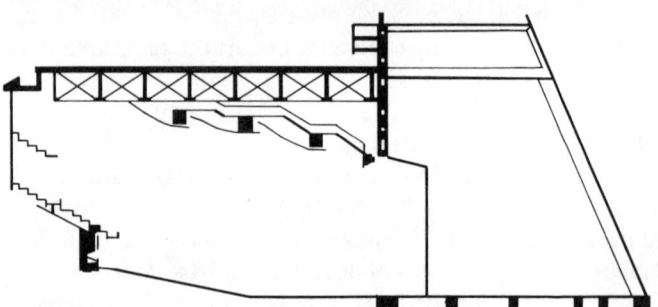

FIG. 5.10.   National Theatre of Guatemala—longitudinal section.

FIG. 5.11.   National Theatre of Guatemala—view towards stage.

trusses and extending the volume of the hall to the upper level of the trusses; and 3, a number of vertical baffles were mounted along radial lines in this extension of the volume.

Proposals 1 and 2 were aimed at increasing the relationship between height and width of the auditorium, to increase the energy of lateral reflections, and 3 had the same purpose as in the Ruben Dario Theatre.

FIG. 5.12.   National Theatre of Guatemala—view towards auditorium.

The proscenium zone was also changed so that the angling of the proscenium side walls was reduced (to 10°) and a movable orchestra shell was designed so that the sides became practically flush with the side walls of the proscenium (Figs. 5.9 and 5.10).

The loudspeaker arrangements had five sound columns, including one centre column above and two on each side of the proscenium frame, strictly for speech purposes. For other purposes there are two cinema units (also suspended above) and 48 single loudspeakers distributed over the auditorium. The sound system also includes a time delay unit with four variable delays. The auditorium has a decorative grill ceiling, of expanded metal, similar to those of the State Theatre and the Ruben Dario Theatre. Views of the theatre are shown in Figs. 5.11, 5.12 and 5.13.

FIG. 5.13.   National Theatre of Guatemala—exterior view.

The opening of the National Theatre took place in the summer of 1979. Final tests were undertaken on that occasion. Values of RT and EDT are stated on the data sheet (p. 91). Values of LE (%) are given in Table 11, Appendix V.

BIBLIOGRAPHY

Subject	Reference
New York State Theatre	Jordan, V. L., *J. Aud. Eng. Soc.*, **13**, 1965, p. 98.

Metropolitan Opera House—data, old House	Beranek, L., *Music, Acoustics and Architecture*. John Wiley and Sons, New York. 1962.
New Metropolitan Opera House	Jordan, V. L., *J.A.S.A.*, **47**, 1970, p. 408.
Ruben Dario National Theatre	Jordan, V. L., *Ingeniera & Arquitectura*, Nov. 1970, pp. 22–25.

DATA FOR NEW YORK STATE THEATRE

Year of completion: 1964
Volume: $19\,500\,\mathrm{m}^3$
Total area: $900\,\mathrm{m}^2$
Number of seats: 2800

RT measured in theatre with 90 % capacity audience:

125	250	500	1 000	2 000	4 000	(Hz)
1·5	1·6	1·6	1·5	1·5	1·3	(s)

General Description

side walls: wood panelling; at orchestra level stepped panelling
rear walls: wood panelling
ceiling: open mesh (two layers) as underceiling; gypsum boards on concrete
floors: hard asphalt compound
seats: upholstered
pit size: $96\,\mathrm{m}^2$

DATA FOR METROPOLITAN OPERA HOUSE, NEW YORK

Year of completion: 1966
Volume: $30\,500\,\mathrm{m}^3$
Total area: $1220\,\mathrm{m}^2$
Number of seats: 3800

RT and EDT measured in the theatre with 80% capacity audience:

	125	250	500	1 000	2 000	4 000	(Hz)
RT	2·25	2·0	1·75	1·8	1·65	1·15	(s)
EDT	2·1	2·1	1·7	1·8	1·7	1·3	(s)

General Description:
side walls: wood panelling
rear walls: wood panelling
ceiling: plaster on lath
floors: vinyl tiles, thin vinyl carpeting in aisles
proscenium walls: plaster on lath
seats: upholstered
pit size: 105 m²

DATA FOR RUBEN DARIO NATIONAL THEATRE, MANAGUA

Year of completion: 1969
Volume: 10 200 m³
Total area: 660 m²
Number of seats: 1400

RT (1/3 octave values) and EDT (octave values), measured in empty hall:

	125	250	500	1 000	2 000	4 000	8 000	(Hz)
RT	1·85	1·85	1·8	1·85	1·7	1·5	0·95	(s)
EDT	—	1·8	2·1	2·1	2·0	1·8	—	(s)

General Description:
side walls: wood panelling
rear walls: wood panelling
ceiling: asbestos panels on concrete, vertically suspended baffles (eternite); grill ceiling as underceiling
floors: parquet
stage enclosure: vinyl panelling
seats: upholstered
platform size: 200 m²
pit size: 84 m²

DATA FOR NATIONAL THEATRE, GUATEMALA C.A.

Year of completion: 1978
Volume: 23 000 m³
Total area: 770 m²
Number of seats: 2050

RT (1/3 octave values) and EDT (1/3 octave values) measured in empty hall:

	125	250	500	1 000	2 000	4 000	(Hz)
RT	2·0	1·9	2·0	2·0	2·0	1·8	(s)
EDT	1·8	2·0	1·7	1·8	1·9	1·7	(s)

General Description:

side walls:	wood panelling, 50 % with slots
rear walls:	wood panelling, with slots
ceiling:	asbestos panels on concrete, vertically suspended baffles (eternite), grill ceiling as underceiling
floors:	marble
stage enclosure:	vinyl panelling
seats:	upholstered
platform size:	280 m²
pit size:	120 m²

Chapter 6

Sydney Opera House

The ambitions of a Labour Prime Minister in the State of New South Wales plus the aspirations of a famous conductor resulted in a challenge to architects all over the world to take part in a competition in 1956, the aim of which was to provide the city of Sydney with an 'Opera House'.

This term, misleadingly applied, could be said to have had an ominous influence on the illustrious project from the beginning. The very words, 'Opera House' conveyed the impression of prestige and vanity and dominated the legitimate aspirations of the friends of musical life in Sydney —a good, modern concert hall to be the future home of the Sydney Symphony Orchestra—aspirations which were shared by the Australian Broadcasting Commission (ABC).

Even in the programme for the competition, the 'major hall' was imagined as a dual purpose hall (capacity around 3000 for concert and 1800 for opera), whereas the 'minor hall' was described as an auditorium for drama and light opera (capacity: 1200). The dual purpose idea, just like the term, 'opera house' assisted in creating an ambiguous situation, where different groups of Sydney society would try to influence the design of the major hall according to their special preferences.

In his approach to the design, the winner of the competition (the Danish architect Jørn Utzon) was obviously influenced, primarily by the simplicity of the Greek amphitheatre but also by some profoundly personal ideas, indeed sculptural conceptions, which were original to the point of being revolutionary in architecture, but which, unfortunately, had absolutely no association with the classical design of concert halls.

In this context it is natural to refer again to the general development in the design of concert halls in this century with the trend away from the

92

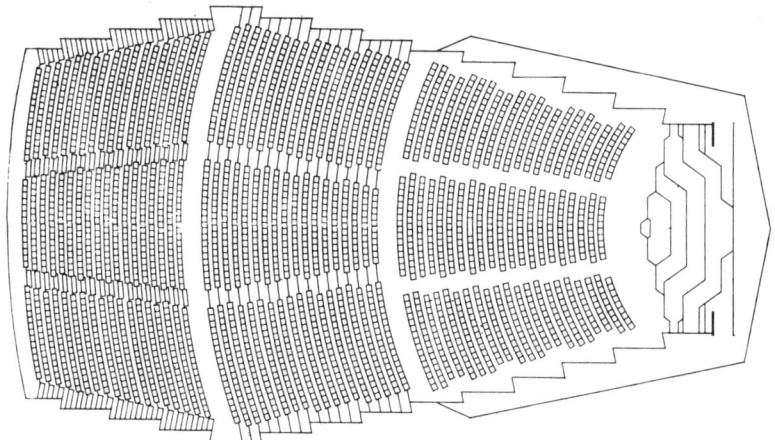

FIG. 6.1.   First design of Major Hall, SOH—plan.

classical (rectangular) shape towards new, sometimes theatrical, shapes, either thought to be useful acoustically or expressing highly personal ideas. Basically, the exterior system of shells, a fascinating concept (concealing the stage towers), was completely unrelated to the interior shape of a 'normal concert hall', and the story of the first design period (the Utzon period, 1956–66) may be interpreted as consisting of several frustrated attempts at trying to relate the exterior shells to the interior boundaries of the halls.

Starting from scratch, an early attempt was made to shape the interior contours of the major hall so that something approximating to 'right angle orientated' boundaries emerged (see Figs. 6.1, 6.2, 6.3 and 6.4). In plan, this was comparatively easy to accomplish by applying the well established principle of stepping the surfaces parallel to the long axis of the hall. In section, particularly in cross section, this was not so easy. In longitudinal

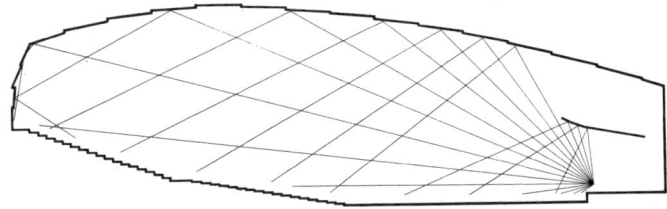

FIG. 6.2.   First design of Major Hall, SOH—longitudinal section, Concert Hall.

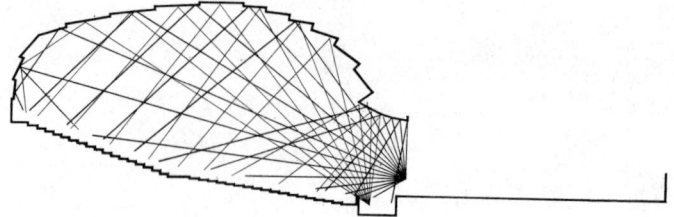

FIG. 6.3.   First design of Major Hall, SOH—longitudinal section, Grand Opera.

section, a gentle curve with steppings was thought to be an acceptable approach. In cross section the close-to-vertical side wall elements were limited in height due to the contours of the shells. Actually, this first design did not account realistically for the limiting demands of the shells.

Without analysing the cooperation between architect and acoustician too minutely, it is fair to say that concerning this first design, the acoustician was responsible for many suggestions.

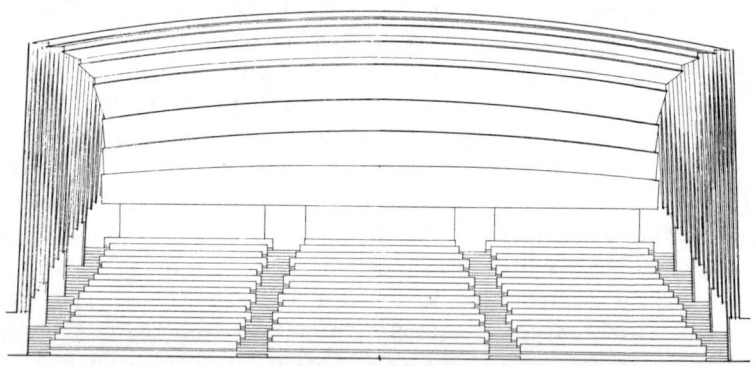

FIG. 6.4.   First design of Major Hall, SOH—cross section.

The construction of the podium and the shells was well advanced (disregarding here many other problems involved) while the design of the interiors passed into a new phase which apparently aimed at a closer approach to the architect's fancies. An extremely beautiful design of the ceiling shape (of 'diamond-like' appearance) would also have been very effective as a sound diffusing ceiling. Unfortunately, the exterior limitations, dictated by the shells, suggested an overall shape of the ceiling which did not leave much space for any side walls, at least no large area of vertically orientated surfaces and this meant that side reflections would be almost nonexistent (Fig. 6.5).

Utzon's third (and final) approach to the solution of the interior design of the major hall signified a clear departure from the original scope with regard to audience capacity (see Fig. 4.7). Abandoned was the promising idea of utilising the immense stage area as seating areas in the concert hall (by

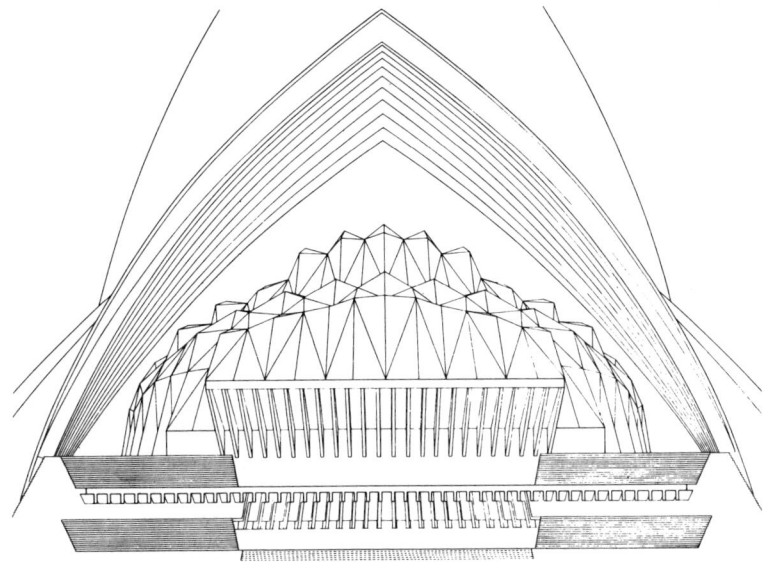

FIG. 6.5. Second design of Major Hall, SOH—cross section.

mechanically closing off the fly tower with sound-reflecting surfaces). The audience seating was pushed backwards and extended laterally having a fixed proscenium frame as permanent limit. This inevitably meant a substantial reduction in the capacity of the hall, and this would have become a major obstacle to the viability of the design, if it had not been for many other controversial problems which were beginning to become serious and which finally hampered the progress of actual construction on the site.

At this point the most unexpected event changed the whole future of the project—the architect resigned. This is not the place to relate details but it is important to stress the significance of the change in administration, since the developments that followed are comprehensible only if one understands the complete break in the continuity of the project and of the construction process. It is not the place, either, to assess the conduct of any of the parties involved nor the justification of their acts, but to give an account of the completion of the job. This includes several phases: the organisation of a new administration, the reprogramming of the whole complex, the final design of several auditoria, their construction and completion.

Meanwhile, pretesting of the third Utzon design of the major hall had taken place. It has already been related (Chapter 4) that testing of a 1/10 scale model of this design, using the criterion steepness (Appendix II), had taken place. According to the results of this testing, the laterally extended seating, with many seats remote from all reflecting surfaces, exhibited low values of steepness. It had also been demonstrated that the suspension of several vertical baffles under the ceiling could to a certain extent improve this situation (see Figs. 4.8, 4.9 and 4.10). However, the events made the third Utzon design and its model obsolete.

The minor hall, also, had undergone several design phases. An early design with a rather eccentric ceiling shape showed the centre part of the ceiling dominated by an upward thrust (Fig. 6.6). The cavity which was created thereby seemed predestined to cause concentrated ceiling reflections, considerably delayed compared with the direct sound. This design was abandoned without being tested in scaled models.

The final design of Utzon for the minor hall had quite another appearance. The curved shapes had now been turned upside down, creating

Fig. 6.6.   Second design of Minor Hall, SOH—longitudinal section.

a ceiling with large convex arches (a variation on his final design of the major hall which had many sharp, plane breaks in the ceiling).

It is necessary at this point to mention that the final designs of Utzon for both the major and minor halls were the results of his cooperation with other consultants. Over a few years the situation concerning acoustic consultancy had developed from a simple request of wanting 'a second opinion' to the ambiguity of uncertain responsibilities, and missed opportunities of cooperation between the parties concerned. This unhappy situation still existed at the time of the architect's resignation. After the reshuffle, the new team of architects (Hall, Todd & Littlemore), represented by the design architect (Peter Hall) submitted their proposals for the team of consultants to the construction authority (Minister for Public Works, Sir Davis Hughes), who subsequently approved their selection. The author remained on the job as *the* acoustic consultant and a consultant theatre architect (the late Ben Schlanger of New York) was appointed and joined the team.

A decade had gone by since the competition, and the programme could no longer be considered as really up-to-date. On the suggestion of the design architect the first task to be undertaken was to revise the programme for the complex by means of a completely fresh approach to the question: What does the city of Sydney need in terms of a cultural centre? A working group consisting of the design architect and the consultants on theatre architecture and acoustics was established and started out to investigate this question which also had to take into consideration the (related) practical question: What, actually, can be accommodated 'under the shells'? As a matter of fact, the shift in administration practically coincided with the situation where the exteriors of the shells were close to completion but where the interiors were left practically empty. The most basic problem to be attacked was the question of the application of the major hall. The original concept of a dual purpose hall did not seem justified in view of the pressing need of the city for a modern concert hall and the much less urgent need for a large scale opera theatre. Realising how the dual purpose concept had haunted this project from the very beginning and how much better a single purpose hall could be designed to fulfil its purpose, a submission was made to the minister that the major hall should be purely a concert hall. It was also proposed that the scope of the minor hall should be revised (and capacity increased) so that it became *the* opera theatre of the complex. It is most important here to report these two major recommendations, plus a third which added another very useful unit to the complex, the orchestra rehearsal and recording studio. In the original scheme there was no space for an adequately sized orchestra rehearsal hall, but when the stage facilities were abandoned (which incidentally meant the dismantling of a fly tower which had already been constructed in the major hall) a huge understage area became available and provided the opportunity of including a spaciously sized rehearsal hall which was also equipped to function as a recording studio.

Further recommendations included a drama theatre with a capacity of 550 seats (originally intended for an experimental theatre) and a music room of 400 seats (of which the scope oscillated between chamber music and arts cinema). The total number of seats in the complex was to increase considerably due to these recommendations, also taking into account that the capacity of the minor hall (now the opera theatre) was increased from 1200 to 1500 plus.

Serious consideration had been given to the inclusion in the recommendations of a change of name of the complex from *Sydney Opera House* to *Sydney Arts Centre* which would have been more in accordance with the spirit of the revised programme, but the idea was dropped as unrealistic.

The design of the most important unit, the concert hall, had to be approached practically from scratch. No preconceived ideas of layout or interior design features had to be accepted but only the basic limitations of the shell system. After a few sketchy proposals, a layout in plan evolved which took advantage of the full length available and which reserved the perimeter areas for the approaches to the auditorium.

At an early stage it was decided that too large distances between performers and audience should be avoided and rather to accept that some of the audience would face the conductor. This was implemented by pushing the orchestra platform away from the extreme end of the auditorium, and using some of the reversed seating for choir seats. The main areas of seating would be orchestra seating with a slight rake, continuing into terraced seating (two terraces, but no balconies). The platform and the orchestra seating would be surrounded at four sides with boxes and terraced seating (Fig. 6.7).

The first design for the ceiling had some (superficial) similarity to the 'reversed' arches of Utzon's minor hall design. In this design, however, they were shaped like catenaries (structurally, a suspended construction), which were also conceived to emphasise a very diffuse sound distribution from the orchestra platform towards the audience seating, in both directions. It was agreed that this design had to be tested by using scaled models (1/10) and models were to be tested in a basement area available at the site. The main

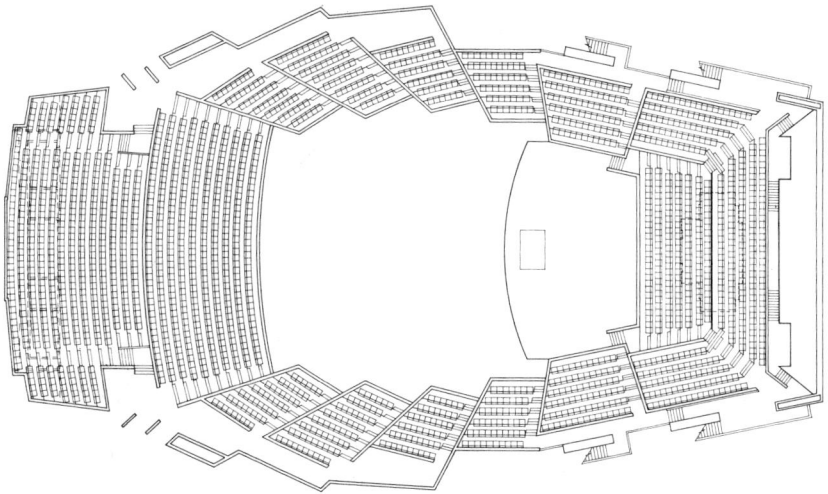

FIG. 6.7. Design of the Concert Hall—terrace level.

criterion to be applied would be values of EDT (and II) after an adjustment of RT had taken place. The audience was simulated by pieces of neoprene with cardboard heads. Another feature of this first design must be mentioned—the side walls which in the recent Utzon designs had been almost non-existent due to the shape of the shells. Since vertical elements at the sides were essential for early lateral reflections, it was suggested that the side walls should be stepped, approaching the contour of the shells. The horizontal elements constituted the catenaries in the longitudinal section. Owing to the large dimensions of the steps and the arches this ceiling shape was rather impressive, not to say somewhat overwhelming (Fig. 6.8).

The testing of EDT at different locations in the model indicated that in some areas of the orchestra seating the risk of getting deficiencies could not be neglected (comparing values of EDT with value of RT). This might signify that the direct sound (plus the ceiling reflection) dominated the early reverberant sound and that the contribution of the early lateral reflections was negligible. Adding suspended reflectors above the stage area did not change this situation.

A radical change in the design seemed necessary, and it was concluded that the ceiling and the side walls would have to be modified. In order to reduce its influence it was proposed that the ceiling should be moved upwards as much as possible and straightened out such that it became

FIG. 6.8.   Concert Hall model—view of first ceiling design.

almost flat, with some stepping. It was proposed that the distance between the side walls should be reduced so that the side boxes became located under soffits and the side walls became continuous vertical boundaries, above the soffit level.

Including several other modifications, the second (and final) design of the concert hall emerged. The layout in plan was modified but retained the gross shape, of a double spade, and the ceiling had a series of large ribs radiating from a crown piece situated above the platform (Figs. 6.9 and 6.10). This design also was built as a scaled model in 1/10 and tested for EDT and II. It soon became evident that the general trend now was that EDT values were in the main larger than RT values and that the deficient values of EDT in the orchestra seating area had disappeared.

Generally speaking, this might be ascribed to the modified design of the ceiling and the side walls which, incidentally, could be interpreted as changing the general shape somewhat in the direction of a 'classical concert hall'.

The development of the design of the Sydney Concert Hall may probably be regarded as the first major example of a modern concert hall where the general shape, plus many details, were radically influenced by thorough scale model testing.

The design of the opera theatre was also influenced considerably by a similar procedure of model testing. The first design of this theatre (after the

FIG. 6.9. Concert Hall model—view towards the stage of second ceiling design.

FIG. 6.10. Concert Hall model—view towards the rear of the second ceiling design.

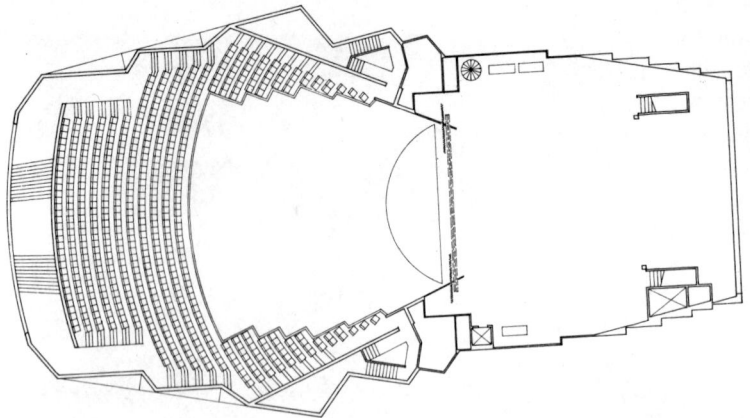

FIG. 6.11.   First design of the Opera Theatre—plan.

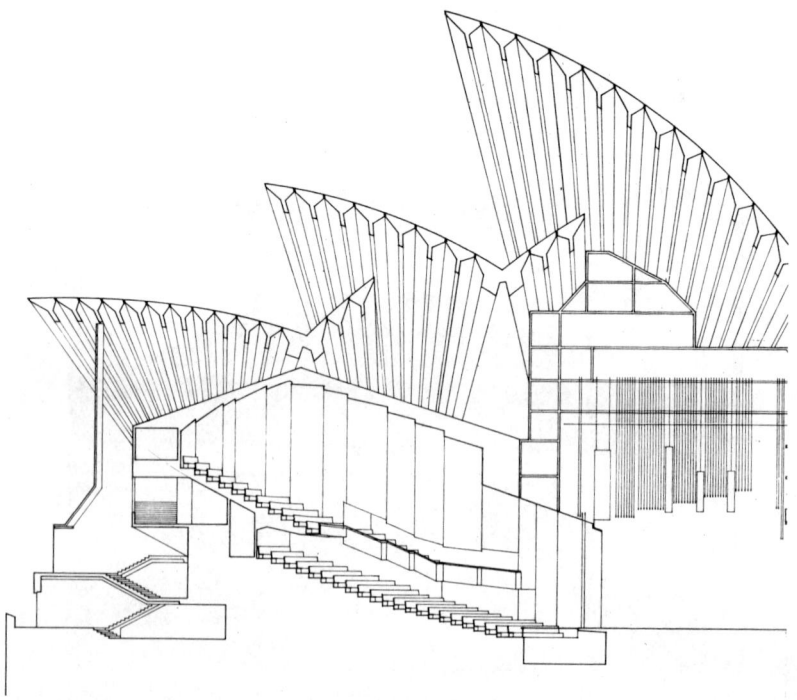

FIG. 6.12.   First design of the Opera Theatre—section.

change of design architect) had a layout with orchestra seating, balcony and one row of side balconies. This increased the capacity from the original figure of 1200 to more than 1500. The ceiling shape was almost flat and sloping and the ceiling height was very moderate. The proscenium was framed by two vertical (diverging) reflectors and an overhead reflector protruding into the auditorium (Figs. 6.11 and 6.12).

By model testing this design it was established, somewhat surprisingly, that EDT values were, in this case, considerably higher than RT values, in fact so much higher that it was suspected it might adversely affect articulation.

The problem of reducing EDT values was experimentally tested by several different design changes which were all simulated in the model. It became evident thereby, that by moving the ceiling upwards and by eliminating the protruding part of the horizontal proscenium reflector the EDT values would approach the average RT value.

It also became evident that the values of II, calculated in the usual manner from measured EDT values, improved. What was considered particularly important was that values measured with the sound source in the pit approached values measured with the sound source located on the stage.

This resulted in a final design with a high ceiling and with three rows of side loges (keeping the total number of seats high but allowing some of the seats in the loges to have bad sight lines, see Fig. 6.13).

The different stages of development in the design of the opera theatre were evaluated according to their values of inversion index. The importance of this concept has been mentioned earlier but it showed itself to be particularly useful in connection with the assessment of design stages of the opera theatre. During the measurement of EDT in the various models, the

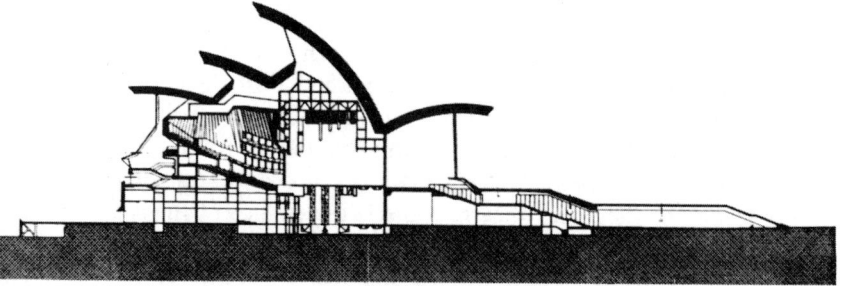

FIG. 6.13.   Final design of the Opera Theatre—section.

location of the sound source was interchanged between the stage area and the orchestra pit area. In each case some microphone locations were included in the same area as the sound source. Thus, on calculating II it became necessary to distinguish between the stage case and the pit case and to establish the individual values of both cases for each design stage. This led to an instructive way of following the variation of II with design change (Appendix III, Table 7A). In the early stages of design, with the flat ceiling and the protruding reflector above the proscenium, the values of II indicated a considerable discrepancy between the stage case and the pit case which might be interpreted as showing a correspondingly poor balance between acoustical conditions for the singers and for the orchestra.

Gradually, with the change of design we notice a change in the values of II which effectively reduce this discrepancy. It is also noted that this is not specifically a question of the frequency range although if one takes more octave values into consideration the discrepancy between conditions for the stage and for the pit decreases somewhat.

With a total number of seats of more than 1500 this project for the opera theatre was considered adequate as a solution to the opera demands of the city of Sydney, even if it meant that 'Grand Opera' would have to be performed on a stage area which was not exactly in accordance with modern trends in stage design, i.e., with plenty of wing space. It could also be argued that if the stage of the major hall had been retained, the opera performances there would have had precisely the same problems with limited wing space. This was a direct consequence of two facts: the geometry of the shell systems and the placement of the two halls side by side on the narrow peninsula of Bennelong Point (a feature which was specific for the Utzon project and which had indisputable aesthetic and visual advantages, but severe practical limitations).

It is extremely difficult to be fair when trying to evaluate the intentions of the original programme compared with the conclusions of the revised programme, especially for one who has been involved in suggesting many of the changes. It is gratifying to be in a position to quote an independent source of judgement. John Yeomans (author and journalist) in his book on the opera house ('The Other Taj Mahal') has made some remarkable comments, including his first reaction to the revised programme—'I had not expected anything so daring, practical and deeply satisfying'.

Returning to the opera theatre project, the change of the scope from drama/intimate opera to full scale opera also meant enlarging the pit. By moving the revolving stage backwards and increasing the depth of the pit this enlargement made it feasible to increase the number of musicians to a maximum of 70. The strong curvature of the pit had to be accepted even if the shape of the pit area was by no means ideal.

Of the many problems involved in the design of auditoria the problem of getting a satisfactory low ambient noise level in the audience and performing areas had to receive careful consideration. In the case of the Sydney Opera House this problem had many difficult aspects. The harbour noises did not seem to present too difficult a problem, although train noise from the nearby harbour bridge and noise from fog horns as well as passing passenger ferries had to be considered.

The most disturbing of the harbour noises was the hooting of the big overseas passenger liners when leaving Sydney. This was a very low pitched sound of considerable intensity, which incidentally was first experienced by the consultants at one of the early programme sessions in 1966. On that occasion it was taped and analysed.

The exterior boundaries of the opera house were to be large glazed surfaces closing the openings of the shells, supplemented with some louvre walls. The inner boundaries between foyer spaces and auditoria were to be relatively light constructions because of the load restrictions of the shells.

In the first design period (of Utzon) certain criteria for the sound reduction were established, but a decision on the design of constructions was never reached during that period. When the final design started, the problem of achieving a reasonably good sound reduction was studied in detail. The total path of sound transmission, from the exterior via the glass surfaces, the foyers and the interior walls and ceiling structure, had to be considered. It soon became obvious that it would not be possible to achieve exactly the same standard of sound reduction for both the large halls since space limitations in the case of the opera theatre meant that this auditorium could not have a completely independent inner roof construction.

One of the first facts which had to be established was the magnitude of the sound reduction which could be obtained through the glazed surfaces between the exterior and the foyer spaces. A series of tests were conducted at the sound laboratories of CSIRO (Commonwealth Scientific and Industrial Research Organisation) on various composite sheets of glazing and the sound reduction which could be achieved with a laminated sheet of 18 mm thickness was established. Originally, it was hoped to have double glazing, but excessive costs made this impossible.

The next step was to establish the design for the interior layers between foyer spaces and auditoria. After various considerations the following design was adopted: a supported steel mesh (suspended from the shells) was to be sprayed with a layer of concrete (5 cm), thereby creating a complete 'cocoon' which joined with the walls of the auditorium (concrete wall—minimum thickness, 10 cm). The inside of the ceiling would carry the panel construction with panels which were a plywood–plaster combination. The panels had been tested and developed especially to provide a resilient membrane, with a low mass/stiffness resonance and a low absorption coefficient at low frequencies. This interior panelling also contributed to the sound reduction efficiency of the construction. The shell structure plus the interior 'cocoon' with the panelling could be estimated to have a total sound reduction of probably more than 70 dB. The passage via glass facade–foyer space–'cocoon' (plus panelling) could not be expected to reach the same reduction figure. As a (modest) supplement to the total attenuation along this route it was suggested, and accepted, that the upper side of the 'cocoon' should be treated with a heavy layer of mineral wool, thus adding to the sound absorption of the space above the inner ceiling.

An actual testing of the sound isolation achieved between the exterior and interior of the halls was not possible to arrange, but it was possible to test the ambient noise level in the auditoria under different exterior conditions. To establish an extreme situation of exterior noise level it was arranged to have a helicopter (from the Defence department) hovering above the shells at approximately constant height (60 m above sea level), while the resulting ambient noise level was recorded (in octave bands) inside the auditoria. Afterwards the ambient noise level without the helicopter was also recorded. The result of this test for the concert hall is indicated in Fig. 6.14. The actual noise level inside the concert hall caused by the helicopter is only distinguishable from the ambient noise (i.e., without the helicopter) in the octave bands 31·5–500 Hz and the noise level per octave does not exceed N15 in any of these octaves.

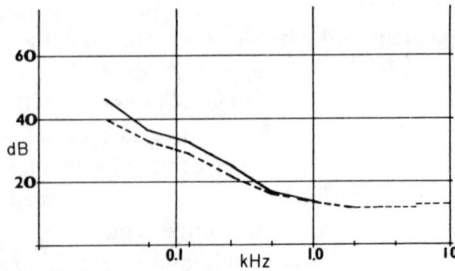

Fig. 6.14.   Noise levels in the Concert Hall due to helicopter outside the building (60 m above sea level). Level (dB) v. frequency (kHz). ——, noise level with helicopter present; –––––, ambient noise level.

By the end of 1972 the two large auditoria of the opera house were close to completion and it was agreed between the parties concerned that test performances admitting full audiences should be arranged. It was also agreed that the test performances would include acoustical testing involving the firing of gunshots from the stage area and simultaneous taping of the impulse response at a number of locations in the halls, both with and without an audience. The acoustical testing would also include the 'classical' method which uses as sound source an orchestra playing the first bars of the 'Coriolanus' overture (by Beethoven). These opening bars include some fortissimi, abruptly interrupted by the conductor. (This method goes back to 1935 where it was used in the Old Philharmonie of Berlin. It has since been used by the author at various inaugurations.) When conductor and orchestra cooperate closely to achieve an instantaneous interruption of all instruments the result is a genuine reverberation process, which when analysed by octave filtering and recording, can establish the values of RT, with an audience present. Of course, it is assumed that the individual decay process of all instruments is negligible compared with the decay process of the hall. This should be checked particularly in the case of tympani and drums.

Of the various parameters analysed, the RT, EDT and II values were studied under different circumstances. The RT v. frequency characteristic in the concert hall received special attention during the testing period because it had been arranged that certain areas of the wall panelling (along the perimeter walls) could be changed in absorption characteristics if desired. A wooden panelling with strips, and slots between the strips, acted as a resonant absorbing panel with frequency maxima in the range 450–500 Hz. By covering some (or all) of the slots with profiled neoprene (fitted into the slots) this resonant frequency, and absorption maximum, could be reduced to 200 Hz.

The first RT v. frequency curve measured was flat except for a local increase of the RT value at around 200 Hz. By closing four out of five consecutive slots per panel unit it should theoretically be feasible to displace the maximum absorption enough to equalise the RT v. frequency curve. This procedure was recommended and carried out after the test concerts. By measurements following the adjustment the change was checked and it was noticed that the equalisation had been obtained (Fig. 6.15).

Even during model testing of the concert hall the influence of reflectors above the platform area had been investigated and it had been found that such reflectors had an influence on the value of inversion index. In the model they had been located at the level of the side box soffits or at the maximum level possible under the ceiling. The total area of reflectors had corresponded to 40–50% of the area occupied and the reflectors were circular (slightly convex at the underside) with diameters corresponding to 1–2 m, in full scale.

When considering the actual design of reflectors in the concert hall different shapes were studied. The toroidal shape adopted occupies a

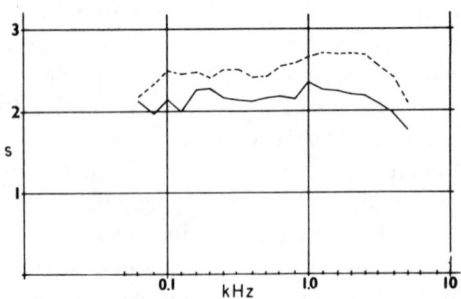

FIG. 6.15. Concert Hall—RT (s) v. frequency (kHz) curves. – – – –, empty hall; ———, capacity audience.

smaller percentage of the total area but they are assumed generally to be more diffusing, and also somewhat more diffusing in horizontal directions. They are mechanically operated so that the level may be adjusted at will.

On the occasion of the test concerts they occupied three different positions: 1, at the level of side box soffits, 8 m above the stage floor; 2, 10·5 m above the stage floor; and 3, just below the crown piece ceiling, 20 m above the stage floor. Variations of the inversion index were investigated for these three positions (Appendix III, Table 8B). Even if variations are comparatively small it is indicated (from the measurements with capacity audience) that position 2 was preferable and this position was adopted as the standard position.

By the time of the test concerts the general impression was that the acoustics of the concert hall would be in accordance with the objective test results: the character of the musical sound would be bright, brilliant and with a considerable dynamic range.

First impressions do not always last, but in this case it became evident that the season of inauguration (from September 1973) confirmed the first impressions and both vocal and instrumental soloists together with conductors expressed their appreciation. Birgit Nielsson, who performed at the opening night, was enthusiastic in her comments.

In the opinion of the author quite a few objective features of this hall may be of importance for the subjective impressions. To list the more important:

1. Considerable areas of reflecting surfaces surrounding the orchestra (including those of the suspended reflectors), providing the musicians with early reflections.
2. The shape of the hall which constitutes increasingly larger reflective zones, as the sound is gradually propagated from the platform (i.e., platform area, orchestra area, 1st terrace, 2nd terrace).
3. Considerable diffusion from the many obstacles on the main ceiling (e.g., ribs and boxes for ventilation inlets).
4. RT (or rather EDT) v. frequency dependence where the important high frequency reverberance is well preserved.
5. Considerable contribution of lateral reflections from the vertical side walls.

The test concert in the opera theatre was also associated with objective testing of the parameters RT, EDT and II. They were measured and calculated for the empty as well as the occupied auditorium—the audience had to endure the gun shots but were warned to protect their ears while the shots were fired.

One of the problems envisaged for the opera theatre was whether it would be possible to get sufficient reverberation with an audience present, since the volume per seat was rather low. It was thought advisable to apply leather upholstery to the seats, this having less absorption at high frequencies when unoccupied. This means that whenever seats are free during performances there will be a slight increase in the value of RT, especially at the higher frequencies.

The measurements of RT show this very clearly (Fig. 6.16). With capacity audience the value of RT is close to 1·4 s, which is somewhat more than was anticipated and quite an adequate value for this size of an opera auditorium. This value is nearly constant over a considerable frequency range (100–2000 Hz).

When considering the values of II it is rather astonishing to notice the close agreement between values measured in the auditorium and those measured in the model (see Appendix III, Tables 7A and 7B). The individual positions used in both auditoria for taping the impulse response were in fact the same as those used in the models, but when comparing, for example, values of EDT, it was not possible to see any close correlation. It looks as if variations of EDT, generally speaking, are too low to be indicative of 'local' acoustical conditions with the exception of those cases where this variation

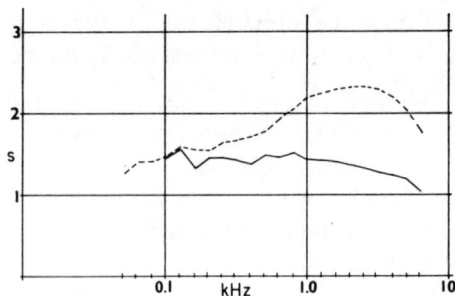

Fig. 6.16.   Opera Theatre—RT (s) v. frequency (kHz) curves. – – – –, empty hall;
——, capacity audience.

is very pronounced (20–30% or more). In this respect the parameter steepness, as we have seen earlier (Chapter 4), may be more susceptible to 'local' acoustical conditions.

The test concert in the opera theatre, as well as the opening performances (in September–October 1973) also received favourable comments from the music critics. Some problems remained to be solved; the sound levels in the pit during fortissimi passages of the orchestra were sometimes excessively loud.

It was recommended that the low frequency absorption panelling which had already been installed in the pit (on the walls) should be supplemented with similar treatments on the underside of the pit ceiling.

The problem of the size of the pit may come up again and there are still certain possibilities of increasing the pit area. (Actually, an increase of the pit area was achieved during 1978.) Leaving the final judgement on all questions, i.e., acoustics, functions and facilities, to the users of the auditoria, it is appropriate here to conclude that under the given geometrical space limitations (below the shells) it is believed that the two main functions of symphony concerts in the concert hall and opera performances in the opera theatre have been well adapted to the building complex. Figures 6.17 and 6.18 show views of the concert hall and Fig. 6.19 is a view of the opera theatre.

The second priority of the concert hall is congress (requiring speech facilities) and at an early stage it was realised that a high quality public address system had to be included in the design. It is not easy to solve the problem of producing satisfactory speech articulation for 2700 persons in a concert hall with more than 2 s reverberation time. In some instances it has been considered necessary to apply a system of distributed loudspeakers

FIG. 6.17. Concert Hall—view towards the organ.

FIG. 6.18. Concert Hall—view towards the rear.

FIG. 6.19.   Opera Theatre—view towards the stage.

(using appropriate time delays). However, to preserve the natural quality of the voice and natural orientation of the audience towards the speaker, it is preferable, if at all possible, to rely on one central loudspeaker arrangement located at the stage, not too far from the speaker's rostrum.

There are two main solutions to the problem of distribution pattern from a central speaker arrangement: horn speakers or column speakers. Neither type is ideal, but the principle of column speakers was preferred on the basis of experience elsewhere. Directional distribution is required to reduce feedback possibilities and principally the column speaker has the advantage that it can be designed to have a very high directivity (in the vertical plane) but there are definite limitations with regard to the frequency range within which the same polar diagram can be achieved. Towards low frequencies the limit is defined by the length of the column and towards the high frequencies the limit is associated with the single cone speaker's directional properties. In order to maintain a constant opening angle over a large frequency range the column length must be considerable and at the high frequencies the column must be 'tapered' towards the extremities, either electrically by gradually reducing the output to the units near the ends or acoustically by screening off some of the output from the same units.

Both methods were tried out experimentally on two lengths of column (4 m and 1·5 m) and the columns for the concert hall were also made with electrical and acoustical tapering respectively. Preliminary testing of both methods in the hall led to the conclusion that electrical tapering produced the best result and so it was chosen for the final installation of columns. (Amalgamated Wireless of Australia (AWA) participated in testing and was active in developing a new method of electrical tapering.)

In the final installation four columns were included: one column (4 m long) suspended above the stage and orientated towards the rear of the hall; one column (1·5 m long) likewise suspended and orientated towards the seating below the organ; and two columns (1·5 m long) mounted on carriages and located at the front of the stage, left and right. (The suspended units are shown in Fig. 6.17 which is a view of the concert hall towards the organ. Figure 6.18 is a view of the concert hall towards the rear.)

The polar distribution pattern in the vertical plane for the (4 m) unit is indicated for the octave frequencies 250–8000 Hz in Fig. 6.20. The zero to 90° ratio is 10 dB or better and the opening angle (except at 250 Hz) is ±10–12°. A satisfactory speech level of 80 dB was achieved at all seats in the auditorium.

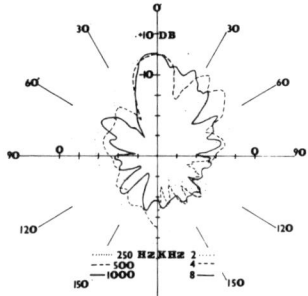

FIG. 6.20.   Concert Hall, 4 m column speaker; polar patterns in vertical plane.

For auxiliary purposes, like entertainment and film, supplementary loudspeaker arrangements with cinema speakers were installed, but it is fair to say that modern groups, electrical instruments, individual speaker systems and very high levels of beat and rock, television shows etc. were *not* envisaged as invaders of the concert hall, neither was their impact foreseen. It is an understatement to say that this situation created problems. The request that all invaders should use the concert hall's speaker facilities and refrain from bringing their own was totally unrealistic.

It was strongly recommended that the use of the concert hall for entertainment of the extremely noisy category be restricted to a minimum. This is by no means an unusual situation today: concert hall acoustics are not compatible with the noisiness of artificially reinforced instruments and singers of modern groups, and there is no easy solution to the problem. In some cases, where the priority of symphonic music is not the absolute and exclusive concern, the dual or multipurpose situation may influence the design of the hall and variable (mechanical or electrical) systems may be considered.

The Rehearsal/Recording Studio of the opera house is located in an area which was originally thought of as the understage area of the major hall, a vast area with a height of 3–4 storeys. The original scheme (of Utzon) did not include a proper sized orchestra rehearsal hall, although the original programme did. This huge understage (with a total volume of more than 5000 m³) was, however, no longer needed for its original purpose when the idea of an opera stage in the major hall was discarded, but the area was put to a real adequate use as orchestra rehearsal space with alternative use as a recording studio. The sound isolation problem between this area and the concert hall had to be solved, taking into account that the platform area of the concert hall was located immediately above the rehearsal/recording studio.

On top of the structural slab closing off the rehearsal hall from the concert hall an extra, floating, concrete slab was placed, resting upon neoprene pads of specified static deflection. No installations were permitted to pierce this area and all structural openings were carefully sealed.

The sound isolation was tested when both concert hall and rehearsal hall had been completed. The test indicated an average level difference of between 60 and 70 dB. The exact figure could not be established because the ambient noise level made the readings in the octave bands 2000 and 4000 Hz impossible (Fig. 6.21).

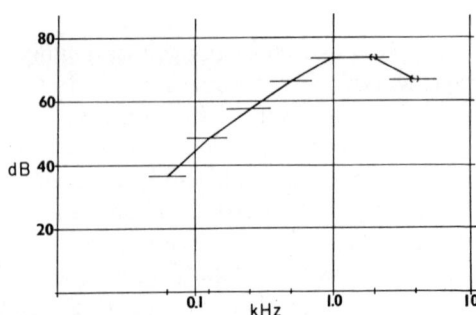

Fig. 6.21. Sound level difference between Concert Hall and Rehearsal Hall (values in brackets were influenced by ambient noise level and were less than or equal to true values). Abscissa—frequency (kHz); ordinate—sound level (dB).

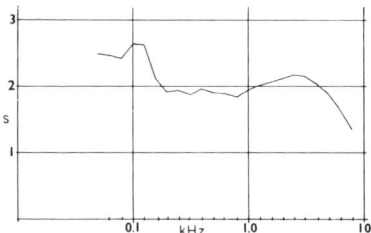

FIG. 6.22.   Rehearsal/Recording Studio—RT (s) v. frequency (kHz) curves.

A request had been made by the orchestra and the ABC that the rehearsal hall should have reverberation characteristics similar to those of the concert hall which meant that RT should be close to 2·0 s. By applying resonance absorbing panels of the type which have their maximum of absorption in the medium frequency range, the frequency dependence could be made to follow the curve described earlier (Chapter 1; for RT v. frequency for the rehearsal/recording studio see Fig. 6.22; for view of the hall see Fig. 6.23).

Originally the rehearsal/recording studio was not meant to have facilities for public attendance but, gradually, it was arranged so that audiences of a few hundred could attend various musical events in the hall.

The Drama Theatre is located at ground level in an area which in the original scheme was planned to house an experimental theatre. By utilising

FIG. 6.23.   Rehearsal/Recording Studio—view.

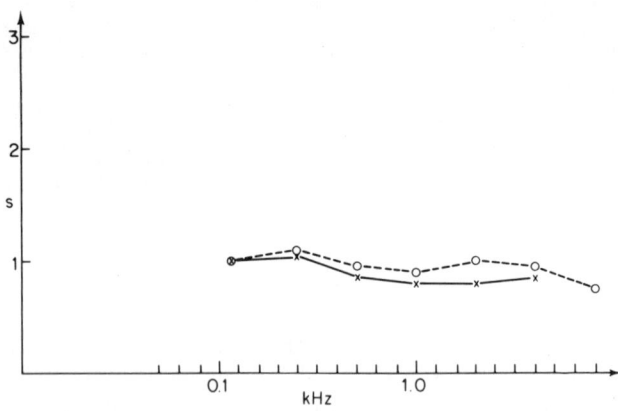

FIG. 6.24. Drama Theatre—RT (s) v. frequency (kHz) curves. —○—, empty theatre (measured); — × —, capacity audience (calculated).

FIG. 6.25. Drama Theatre—view.

the available area it became possible to have sufficient space for an audience of 550 and also for proper stage facilities, including a forestage. RT is close to 0·9 s (Figs. 6.24 and 6.25).

The Music Room, also located at ground level (and also an addition to the programme) was thought to replace the chamber music room of the Utzon design which was only meant to have 250 seats and which was considered to be too small. In the final design phase the scope of this hall varied between two different purposes: *chamber music* and *arts cinema*. After struggling for a number of years with the problems inherent in a dual purpose solution (variable acoustics, movable stage, etc.) the arts cinema proposition was accepted but in a last minute action (by user organisations) it was finally rejected. As the chamber music scope had won this local battle it was imperative that the reverberation should be restored and the absorption reduced. Finally, RT was restored to a value of 0·9 s and the RT v. frequency curve became practically flat. The reports on the acoustics of this hall (which has seating for 400, see Fig. 6.26) were favourable.

The Reception Hall occupies the area where the original scheme had intended to place the chamber music room. It may be used with or without seating space for an audience of 200.

Since the inauguration season of 1973 six years have passed and it is possible now to evaluate the extent to which the opera house has been utilised. The yearly records show astoundingly high percentages of

FIG. 6.26. Music Room—view.

FIG. 6.27.   Exterior view of Sydney Opera House.

attendance and utilisation for most of the halls, especially for the large halls.

Interestingly, the idea to use the concert hall for 'grand opera' has been tried out recently. With a fixed stage setting (covering the organ wall), the singers performing preferably on the platform area and the orchestra players moved out in front of the platform, 'Aida' was performed and the comments indicated a great success from an acoustical point of view.

Thus the old story of the fight between the two purposes of the largest hall of the complex, 'grand opera' and 'symphony concert' has finally come to a conclusion which looks very much like a happy ending. An exterior view of the Opera House is shown in Fig. 6.27.

BIBLIOGRAPHY

Subject	Reference
Sydney Opera House, original design, Utzon	Utzon, J., Sydney National Opera House, 1958, unpublished report.
revised design, Utzon	Utzon, J., Sydney Opera House, 1962, unpublished report.
acoustical design and testing	Jordan, V. L., *Journal and Proceedings of the Royal Society of NSW* **106**, 1973, p. 33.

design, Peter Hall	Hall, P., Todd, L. and Littlemore, D., Sydney Opera House, Stage III, 1968, unpublished report.
evaluation of testing	Jordan, V. L., *Music Room Acoustics*, Royal Swedish Academy 17, Stockholm, 1977.
acoustical design and testing	Jordan, V. L., *Auditorium Acoustics* (ed. R. McKenzie), Applied Science Publishers, 1975, p. 55.
comments on revised programme	Yeomans, J., *The Other Taj Mahal*, Longman Australia Printing Ltd, Camberwell, Victoria, Australia, 1973, p. 176.
comments on acoustics	*Ibid*, p. 216.

DATA FOR CONCERT HALL, SYDNEY OPERA HOUSE

Year of completion: 1973
Volume: 24 500 m³
Total area: 1600 m²
Number of seats: 2690

RT and EDT measured with capacity audience (1/3 octave values):

	125	250	500	1 000	2 000	4 000	(Hz)
RT	2·3	2·3	2·0	2·1	2·1	2·0	(s)
EDT	2·05	2·1	1·95	2·1	2·1	2·0	(s)

General Description:
side walls: wood panelling, some areas with slotted panelling
rear walls: wood panelling, some areas with slotted panelling
ceiling: plaster–plywood combination
floors: parquet
seats: upholstered
platform size: 170 m²

DATA FOR OPERA THEATRE, SYDNEY OPERA HOUSE

Year of completion: 1973
Volume: $8200\,m^3$
Total area: $720\,m^2$
Number of seats: 1547

RT and EDT, measured with capacity audience:

	125	250	500	1000	2000	4000	(Hz)
RT	1·2	1·3	1·2	1·3	1·25	1·2	(s)
EDT	1·25	1·35	1·35	1·3	1·3	1·2	(s)

General Description:
  side walls: wood panelling
  rear walls: wood panelling
  ceiling: plaster–plywood combination
  floors: parquet
  seats: upholstered, leather
  pit size: $91\,m^2$

DATA FOR REHEARSAL/RECORDING STUDIO, SYDNEY OPERA HOUSE

Year of completion: 1973
Volume: $5200\,m^3$
Total area: $520\,m^2$

RT, measured with orchestra present (1/3 octave values):

63	125	250	500	1000	2000	4000	8000	(Hz)
2·5	2·6	1·9	1·9	1·95	2·1	2·0	1·35	(s)

General Description:
  walls: some areas, perforated wood panelling; some areas, concrete
  ceiling: plaster–plywood combination, pyramid shapes
  floor: parquet
  gallery openings: curtains exposed when used as television studio

DATA FOR DRAMA THEATRE, SYDNEY OPERA HOUSE

Year of completion: 1973
Volume: 2100 m³
Total area: 510 m²
Number of seats: 553

RT, measured without audience and RT, calculated with audience (1/3 octave values):

	125	250	500	1 000	2 000	4 000	8 000	(Hz)
RT (empty)	1·0	1·1	0·95	0·9	0·9	0·95	0·75	(s)
RT (audience)	1·0	1·05	0·85	0·8	0·8	0·85	—	(s)

General Description:
side walls: concrete
rear walls: glass, wood panelling
ceiling: aluminium
floors: carpet
seats: upholstered
pit size: 55 m²

DATA FOR MUSIC ROOM, SYDNEY OPERA HOUSE

Year of completion: 1973
Volume: 2100 m³
Total area: 440 m²
Number of seats: 400

RT measured without audience (1/3 octave values):

125	250	500	1 000	2 000	4 000	8 000	(Hz)
—	—	0·9	0·9	1·0	0·9	0·7	(s)

General Description:
walls: wood panelling
ceiling: wood panelling
floor: carpet
seats: upholstered
platform size: 45 m²

Chapter 7

Oslo Concert Hall—Design Development

The Oslo Philharmonic Society had dreamt for many, many years, as far back as the 1920s, that they would have a home for the musical life of the city—indeed a centre for the whole of Norway—a home for the Philharmonic Orchestra with a concert hall of their own.

After the war the dream had to be realised, at long last. An architectural competition was held in 1957 and many entries were received. The municipality of Oslo provided the site and at this point the acoustical problems immediately started, because of the shape of the site. It was, believe it or not, a triangle (Fig. 7.1). Through no wish of the architect, but a problem he had to face, started a design development which, due to many delays, lasted nearly a decade. After this came a long period of construction (starting with an office building) which also lasted nearly a decade. Questions have been raised as to whether the architect (Gösta Åbergh of Stockholm) might have managed to give the large hall inside the complex another shape without reference to the shape of the site. However, one of the reasons why he won the competition may have been the very convincing way in which his interior hall shape fitted the exterior shape of the site.

Questions have also been asked whether or not the acoustical consultant should have taken a strong stand against the architect and should have tried to convince him (and the steering committee) that this shape could be detrimental to the acoustics of the concert hall. The consultant chose to convince the architect (and the committee) that even a gross triangular shape with a considerable size of top angle could form the basis for a concert hall with good acoustics. At the same time, it was strongly recommended that the interior design of the hall should be guided by the results of thorough model investigations. These investigations were continued for

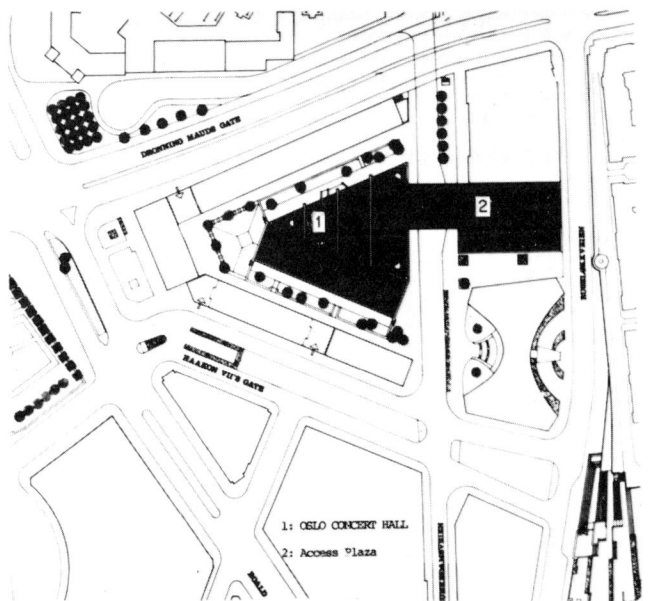

FIG. 7.1.   Oslo Concert Hall—site plan.

much longer than originally expected and involved many more changes in design than anybody had expected at the beginning.

The starting point, i.e., that a triangular shape could be the basis for a good concert hall, had just been proved by the results obtained with the Tivoli Concert Hall of Copenhagen (Chapter 1), although the top angle of that hall was smaller than the angle of the Oslo hall. The fundamental remedy against the divergency applied in the case of the Tivoli Hall was also recommended for incorporation into the design of the Oslo hall—the side walls were to be stepped, with surfaces parallel to each other, and to the long axis of the hall.

The question of ceiling height was discussed earlier (in Chapter 4) as was the problem of the ceiling shape. The ceiling shape was the first problem to be investigated by model research (Chapter 4), and it was concluded that a system of 'beams across' had certain advantages compared with the alternatives, i.e. flat ceiling and 'radial beams'. It was the two criteria, EDT and II in particular, which indicated the advantage of 'beams across' (for values of II see Appendix III, Table 6).

The next problem tackled by model testing was the question of whether the audience seating should be all at one level or whether a rear balcony should be included. The testing method had itself been further developed and now included measurement of EDT values in five octave bands (model frequencies: 1, 2, 4, 8 and 16 kHz). Generally speaking, it was found that numerical values of II did not depend much upon the individual octave band tested. The first balcony design proposed only included seating for 200 (out of a total of 1600 seats) and a seating of 100 choir seats behind the platform (see Fig. 7.2).

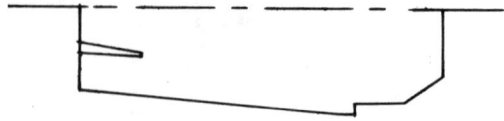

FIG. 7.2.   Oslo Concert Hall—first balcony design.

The criteria applied so far, EDT and II, were not considered to be relevant to a problem which was crucial for the situation investigated: the comparatively long distances from the platform to the rearmost seats combined with the triangular shape. How would this affect the sound level v. distance?

To test this, short pulses from the spark gap were applied and peak levels at different microphone positions were measured with the aid of a rapid level meter of special design. The method thus developed was not quite reliable, but it did indicate that the level drop v. distance might in some cases exceed the distance law for free propagation! This was considered so grave that the revision of the triangular shape continued. By introducing angled walls at the rear of the side walls the extreme corners of the triangle were cut off. The balcony was moved upwards and was transformed into a terrace—the terrace wall then providing reflections for the stalls area.

The level v. distance measurements were indeed affected by these changes and showed a definite improvement. Still, the question of the location of the orchestra required further attention. The obvious idea of moving the orchestra platform somewhat closer to the middle of the hall was now considered. (This idea had already been implemented in the design of the concert hall in Sydney Opera House and it had also influenced the design of other concert halls.) Three (theoretical) possibilities were envisaged: 1, the orchestra being close to the narrow end of the triangle; 2, the orchestra being close to the middle of the hall; 3, the orchestra being close to the wide end of the triangle.

These three possibilities were simulated in the model (see Fig. 7.3) and they were investigated with regard to numerical values of EDT and II. The

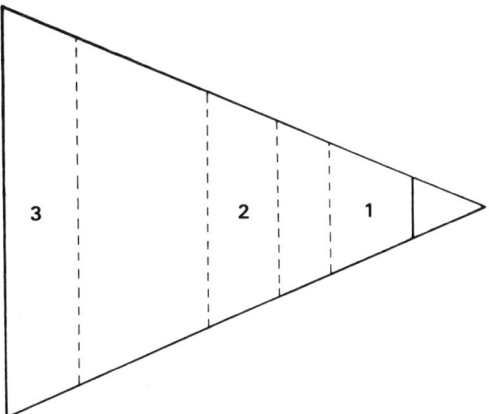

FIG. 7.3.   Oslo Concert Hall model—three (theoretical) possibilities of locating
the orchestra.

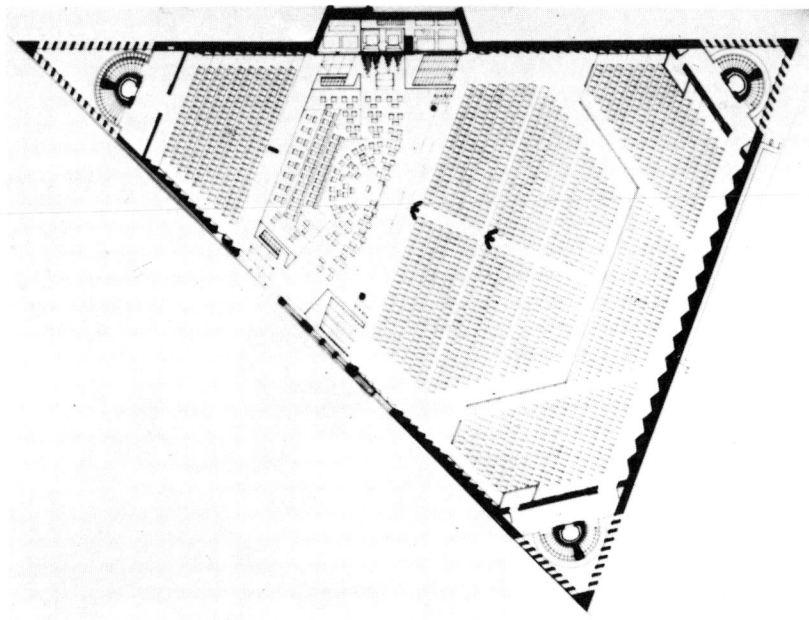

FIG. 7.4.   Oslo Concert Hall, final design—plan.

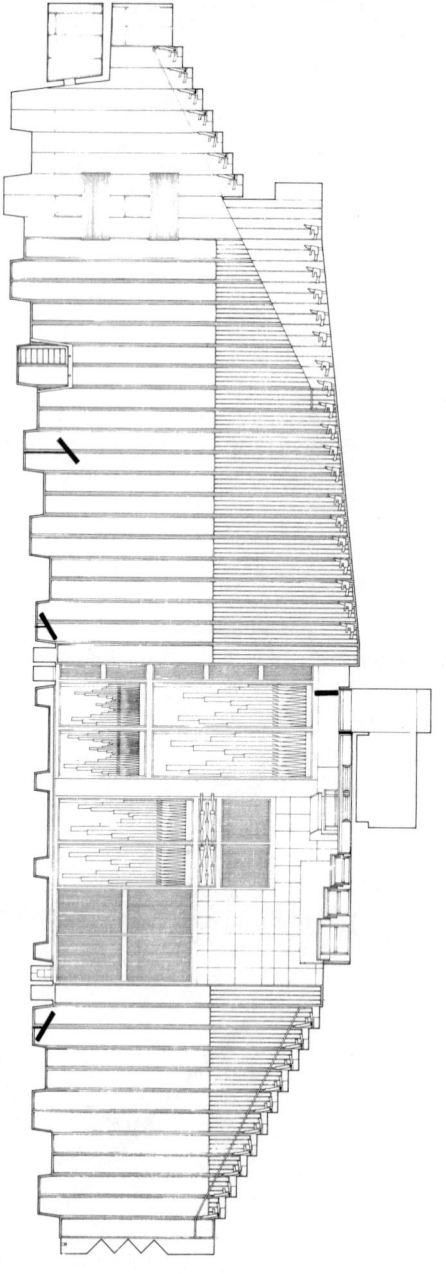

FIG. 7.5.   Oslo Concert Hall, final design—longitudinal section.

result was that values of II were only slightly dependent upon the different situations but the values of EDT showed less variation with location when the orchestra platform was in the middle position. This was a strong indication that such a position would be preferable.

Meanwhile, certain suggestions had been made (by the architect and the steering committee) to combine this change in platform location with the inclusion of a 'small hall' at the narrow end of the triangle. This hall was to be used as a recital hall, screened off from the large hall by a movable curtain, but the two halls were not meant to be used simultaneously. The design of the orchestra platform, including a large area which could be moved up and down by hydraulic lifts, would make it possible to adapt the orientation of the stage towards the small hall (the 'stage terrace') or towards the large hall.

With these features the design of the hall was about to become properly defined and the terraced section at the rear end was further developed by continuing the seating areas at both sides of the terrace down to the level of the stalls, thus integrating the seating areas and also creating more reflecting surfaces which reflected the sound towards the centre area of the stalls. Figures 7.4, 7.5 and 7.6 illustrate the final plan of the hall.

On several occasions it had been discussed whether or not the platform area should have some kind of reflecting surfaces above the musicians. It was decided to try two different types, differing widely in principle. One was the normal type of area reflectors, which in this case were of flat pyramid-like shape (Fig. 7.7) and the other was a kind of 'frame' reflector with mostly vertical surfaces. The latter type was intended to increase the horizontally

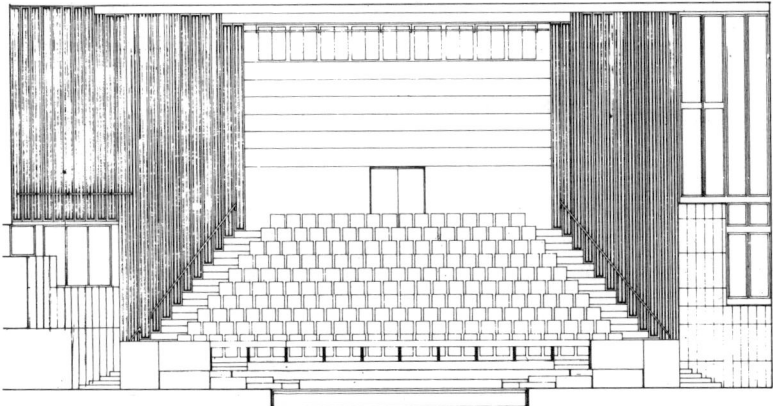

FIG. 7.6.   Oslo Concert Hall, final design—cross section at orchestra platform.

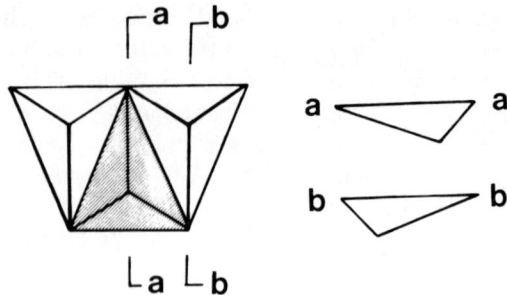

FIG. 7.7.   Oslo Concert Hall, proposed reflectors, flat pyramids.

reflected energy (Fig. 7.8); it was supplemented (for aesthetic reasons) with a number of horizontally orientated planks suspended just below the ceiling.

The analysis of the results of these experiments with reflectors indicated that the best conditions, acoustically, were obtained without any reflectors at all. Apparently, this could be thought of as contradicting the results obtained in the case of the Sydney Concert Hall, but it must be remembered that the Sydney hall has a ceiling height of more than 20 m whereas the Oslo hall has a ceiling height of about 13 m.

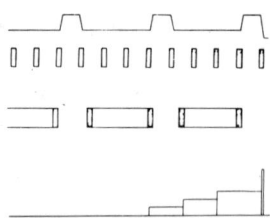

FIG. 7.8.   Oslo Concert Hall, proposed reflectors, 'frame'.

Another question, treated in connection with the reflectors, was whether a flat or raked platform area is acoustically preferable. It appeared that a flat area for the orchestra was advantageous compared with a raked area. However, the Philharmonic Orchestra of Oslo traditionally performed in an amphitheatre arrangement; with the hydraulic, vertically movable sections, both situations could easily be established.

The location of the organ had been the subject for discussion at an early stage. The possibility of a central position behind the orchestra (realised in the case of the Sydney hall) did not go well with the elevated choir seating and was completely ruled out when the 'stage terrace' was introduced. It was eventually decided to have the organ installed to one side of the platform and to disregard the asymmetrical sound distribution and the possible level differences which may result for some of the seats (mainly those in the front part of the hall).

The ample volume per seat of this hall meant that no difficulties were expected with regard to maintaining a value of RT around the 2·0 s mark.

All interior walls and the ceiling were to be panelled—the panel material was a sandwich structure of cement based chipboard and wood veneer—and most of the wall panelling would be solid on risers while the ceiling panelling (also solid) would be supported by a large underceiling of lightweight concrete. Some areas of the wall panelling, however, would be slotted panels, with absorbent backing, which could be closed off with neoprene strips (completely or partly) if it was wished to make smaller changes of reverberation time. This was the same idea as had been applied in the Sydney hall.

During the design changes which had accompanied the model investigations the ceiling had been comparatively little affected except that the resulting profile deviated somewhat from the original simple concept of 'beams across' (Fig. 7.5). In any event it could be regarded as a rather efficient sound diffusing ceiling especially at the lower frequencies.

The recent investigations, by M. R. Schroeder et al., on how to diffuse sound reflections effectively by 'jumps' in a surface (which follow 'a maximum length code') could be thought, indirectly, to throw some light upon the usefulness of 'beamed' or 'coffered' ceilings (and of classical concert halls) with regard to the diffusion of reflections from above. No doubt, a more systematic approach to apply this as a design feature in new halls may be expected.

The model investigations were concluded by several test series of the finalised design and the results led to a number of conclusions on details, e.g., concerning the depth of the rear terrace which was reduced to six or seven rows. Generally, it was concluded that numerical values of EDT and II were all acceptable, especially for the orchestra platform with a flat floor, and concerning the attenuation in level with distance it was concluded that for most test frequencies the values were now below the theoretical values of the distance law.

Some echogrammes showed a rather pronounced delayed reflection (time delay close to 125 ms), which could be interpreted as a reflection off the back wall of the 'stage terrace'. Further diffusion from this wall was obtained by triangular panel breaks.

The systematic tests in the model were followed by a more specific investigation of looking for a possible correlation between numerical values of EDT measured normally with omnidirectional microphones and values of EDT measured with directional receivers. A half-inch condenser microphone which was mounted in the focus area of a parabolic reflector gave a considerable directional sensitivity, especially in the 16 kHz octave band.

The corresponding table of EDT values (average and in different directions, see Appendix III, Table 9) indicates very clearly that the most important contributions to the average value of EDT stem from the lateral-horizontal direction, at least for microphone positions in the stalls area. The conclusion of this investigation would be that, at least in this particular case, the EDT (average value) is affected by the value of EDT measured in the lateral-horizontal direction.

In designing the Oslo Concert Hall considerable effort was expended on securing a satisfactory low ambient noise level of the interior. As protection against outside noise the concrete structural walls were supplemented with gypsum partitions plus thick layers of insulation material. The structural roof was supplemented with an interior ceiling of light-weight concrete suspended under the structural beams. Special care was also taken to

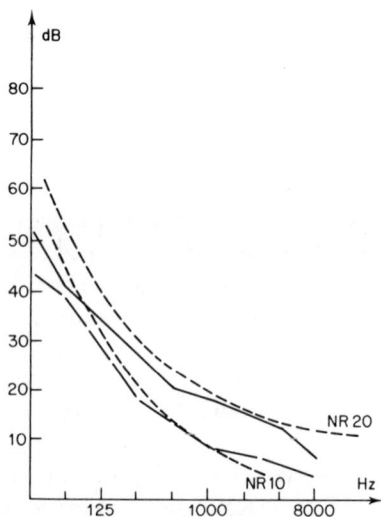

FIG. 7.9.   Oslo Concert Hall—Noise Level, air handling system, normally operated. Noise level: ——, air handling system; – – – –, ambient.

effectively reduce all kinds of ventilation noises, and the air inlets were particularly developed to attain a sufficiently low noise figure at the rated air velocity. The criterion established for the ambient noise level of the hall was N 20 or better. This was accomplished with only minor adjustments. The actual noise levels (per octave) measured with and without the air handling system operating are shown in Fig. 7.9.

Before the official inauguration of the concert hall (views of the hall

FIG. 7.10.   Oslo Concert Hall—view towards terrace.

shown in Figs. 7.10 and 7.11) a test concert was arranged on which occasion the Philharmonic Orchestra of Oslo performed. In connection with the test concert, acoustical testing was also carried out, partially by using the method of recording reverberation processes from the opening bars of the 'Coriolanus' overture and also by firing shots from the stage, recording the impulse response at several locations in the auditorium, and later analysing the tapes.

Besides the normally applied criteria, RT (v. frequency) and EDT at several locations, two new criteria were also analysed. They were *clarity* (C) and *room response* (RR). However, the development which led to their introduction deserves a chapter of its own so the results of their application in the case of the Oslo Concert Hall will be discussed later (Chapter 9).

The RT v. frequency dependence measured with and without an audience is shown in Fig. 7.12. Frankly, it was a surprise to see that the precalculated levelling of RT values as a function of frequency was slightly too much in favour of the high frequencies. Interestingly enough, this frequency characteristic was not subject to any adverse criticism in the

FIG. 7.11.   Oslo Concert Hall—view towards stage.

judgements of either conductors or musicians. The majority were asking for 'a bit more reverberation', without indicating a particular tonal range. The tuning possibility—the slotted panels which could be closed—permitted a slight increase in values of RT over a considerably broad band of frequencies (according to predictions) and it was decided to go ahead with this change. When choosing the areas of panelling where the closing off of the slots should be made, it was decided to choose those areas along the side walls at a low level above the floor. This causes early reflections from these wall areas to be somewhat reinforced over the frequency range in question, approximately 250–4000 Hz.

The resulting RT v. frequency dependence of this change is shown in Fig.

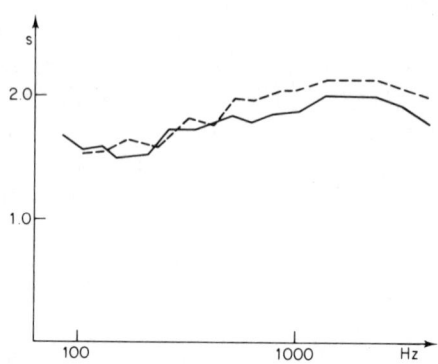

FIG. 7.12.   Oslo Concert Hall—RT (s) v. frequency (Hz) (measured $\frac{1}{3}$ octave values). ——, with capacity audience; – – –, empty hall.

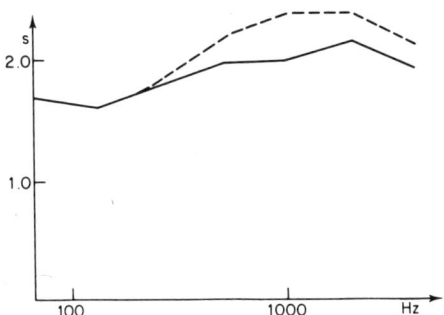

FIG. 7.13. Oslo Concert Hall—RT (s) v. frequency (Hz) ($\frac{1}{1}$ octave values) after correction of panelling. --- RT in empty hall, measured. —— RT with capacity audience, calculated.

7.13 where the calculated curve for the hall with capacity audience is also indicated.

The official opening of the Oslo Concert Hall in March 1977 took place only four weeks after the test concert. The opening programme consisted exclusively of orchestral pieces by Norwegian composers, both new and old. Comments on the acoustics were almost unanimously favourable. Experiences with different types of performances and for a period of more than a year have not indicated any serious defects except for one type of performance, which was listed last on the priority list, i.e. entertainment programmes. The modern groups with their own amplification systems (for individual instruments and singers) have created some particular problems. The situation is to some extent analogous to the one encountered in the concert hall of Sydney. There are no easy solutions to this inherent contradiction between the acoustical conditions which are preferable in a genuine concert hall and the conditions which will favour the modern groups with electronic reinforcement. Obviously these groups are not interested in many reflections and plenty of reverberance but would prefer a rather highly damped interior of the type which is realised in cinemas and television studios. Even if the usage, 'entertainment' was put at the bottom of the list of priorities for the Oslo hall (1, concert; 2, congress; 3, electronic music and 4, entertainment) the experience in practice for the first year of operation has been that number 4 had to be upgraded considerably.

In cooperation with management and architect it was decided to install a series of draperies of soft material—soffits and side pieces—around the stage whenever an 'entertainment' event was on the programme. The necessary overhead installation had already been provided and the amount of drapery can be varied in several steps, corresponding to the various situations on the stage. This arrangement (see Fig. 7.14) has now been in operation since the end of 1977 and the results have been decidedly good. Naturally, the drapery is completely removed whenever the hall is used for symphonic music.

FIG. 7.14.   Oslo Concert Hall—layout of draperies used for pop concerts etc.

The number 2 priority on the list, 'congress', meant that speech quality and articulation should be given serious attention. A high quality sound reinforcement system for this specific purpose had to be provided. It was comparatively easy to supplement the layout for this system with various additional loudspeaker positions which satisfied the composers and performers of 'electronic music'.

To overcome the problems of adequate speech distribution in a reverberant hall it is possible, but no simple matter, to apply a centrally located speaker arrangement (like the one used in the Sydney hall). It is also possible to go to the other extreme of providing a distributed speaker system (like the one used in the Moscow Congress Building, where speakers are located in the back of the seats). In this case the best approach was considered to be the location of a number of sound columns distributed in rows under the ceiling, supplemented with a pair of columns located at the stage (one on each side). A system of this kind will only function properly when combined with an electronic time delay system which delays each column correctly in such a way that all listeners, irrespective of their position in the hall, turn towards the speaker at the rostrum (for arrangement of speakers, see Figs. 7.4 and 7.5).

This speech system is, for other purposes, supplemented with cinema bass units located at the stage so that the 'house system' can actually cope with all normal performance requirements except the groups demanding their own reinforcement systems. Time delay systems of the new 'digital' type have certain inherent problems, which may be specific for the 'first' generation and which are noticeable as high frequency distortion (especially of 's' sounds). This problem also affected the sound system of the Oslo hall in the first season, but has since been overcome by correcting the digital unit.

The Oslo Concert Hall also contains a number of smaller rehearsal rooms but only the recital hall will be described here. This hall, situated in a building adjacent to the concert hall, housing administration, tutorial rooms, etc., has a rectangular shape and the seating area can either be on a flat floor or on a telescopic seating arrangement with a stepped floor. In both cases an upper balcony has some supplementary seats. The number of seats (with the telescopic arrangement) is 300.

The ceiling is covered with pyramid-like shapes to provide a thorough distribution of reflections from above. The walls have triangular breaks

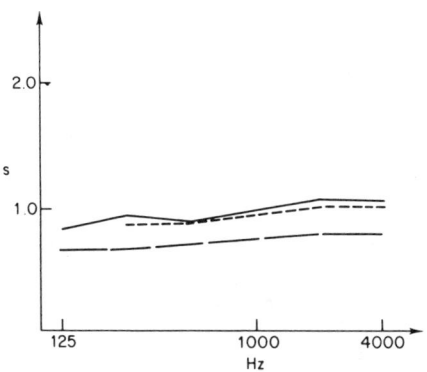

FIG. 7.15.   Oslo Recital Hall—RT and EDT (s) v. frequency (Hz); telescopic arrangement. —— RT measured in empty hall; ---- EDT measured in empty hall; --- RT calculated for hall with capacity audience.

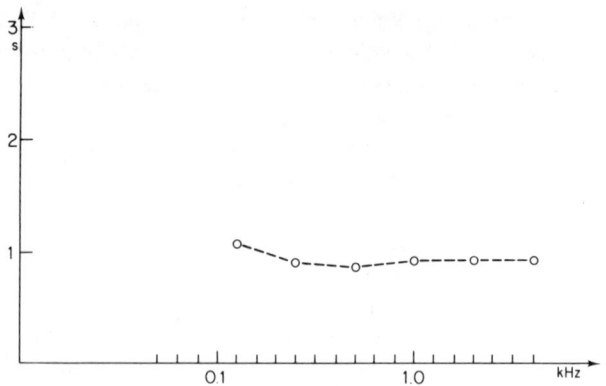

FIG. 7.16.   Oslo Recital Hall—RT v. frequency measured in empty hall. Flat floor without chairs.

FIG. 7.17.   Oslo Recital Hall—view from the platform.

with a few absorbing areas. The platform has a simple elevated wooden floor and may be removed if desired. The testing was limited to the empty hall and the criteria measured were RT and EDT. Both cases, i.e., with and without telescopic arrangement of the seating, were tested. Calculated values of RT with capacity audience are indicated along with the measured values in the empty hall in Figs. 7.15 and 7.16; a view of the recital hall is shown in Fig. 7.17. It is not surprising to find that in a hall of this size values of EDT and RT practically coincide. The average value of RT with the telescopic arrangement (0·7–0·9 s) may be considered rather low for a chamber music hall but still acceptable. The opening concert (May 1977) indicated satisfaction with the acoustical conditions.

Nevertheless, it has been recommended that when used for chamber music, the hall should have a flat floor arrangement, which will result in an RT average value of 0·9 s.

BIBLIOGRAPHY

Subject	Reference
Oslo Concert Hall—design development	Jordan, V. L., *Teknisk Ukeblad*, Ingeniørforlaget A/S, Oslo, 1977.
Oslo Concert Hall	Åbergh, G., *Oslo Konsert Hus*, Oslo, 1978 (unpublished report).
Diffusing the sound by 'jumps'	Schroeder, M. R., *J.A.S.A.*, **57**, 1975. pp. 149–150. Gerlach, R. E. and Schroeder, M. R., DAGA '76, VDI Verlag, Berlin-West. pp. 255–258.

DATA FOR OSLO CONCERT HALL

Year of completion: 1977
Volume: 18 570 m³
Total area: 1555 m²
Number of seats: 1700

EDT and RT, measured in empty hall and RT measured with capacity audience:

	63	125	250	500	1 000	2 000	4 000	8 000	(Hz)
EDT (empty)	1·3	1·5	1·7	1·9	2·1	2·3	2·0	1·3	(s)
RT (empty)	—	1·6	1·8	2·0	2·05	2·15	2·0	—	(s)
RT (audience)	—	1·55	1·7	1·8	1·9	2·0	1·8	—	(s)

General Description:

side walls:	veneered chipboard (cement base); some areas with slots
rear walls:	veneered chipboard (cement base); no slots, triangular diffusers
ceiling:	veneered chipboard (cement base); beams across
floors:	parquet
seats:	upholstered
platform size:	275 m²

DATA FOR OSLO RECITAL HALL

Year of completion: 1977
Volume: 1500 m³
Total area: 250 m² (balcony: 75 m²)
Number of seats: 300

RT, measured in empty hall and calculated for capacity audience:

	125	250	500	1 000	2 000	4 000	(Hz)
(empty)	0·8	0·9	0·85	0·9	1·05	1·05	(s)
(audience)	0·65	0·65	0·7	0·75	0·8	0·8	(s)
(for audience on flat floor)	1·05	0·9	0·85	0·9	0·9	0·9	(s)

General Description:

walls:	veneered chipboard (cement base); some areas perforated
ceiling:	veneered chipboard pyramids on cement base
floor:	parquet
seats:	telescopic arrangement, upholstered (or flat floor)
platform size:	30 m²

Chapter 8

Multiform and Multipurpose Halls

The modern development of assembly halls has led to the concept of 'Multipurpose' which means using the same auditorium for a variety of purposes. The specific development of the theatre, however, has led to another concept, that of 'Multiform' which means using the same volume for a variety of different shapes of theatre. This concept of multiform also has influence upon the acoustical conditions of the space, especially so in those cases where larger halls may have changing shapes.

The development from the Greek amphitheatre with the actors located in a central area and the audience seated in a semicircular, sloping area around the centre, to the Baroque proscenium theatre (Chapter 2) may be regarded as one continuous development which resulted in the 'shoebox stage' where actors' movements are restricted by the proscenium frame. Exceptions, of course, are the Elizabethan theatre, where the actors move around in an oblong stage area, surrounded by an audience, and also the Japanese Kabuko theatre where the actors have access to passages through and above the audience seating area.

This century has seen many tendencies to involve the audience more intimately in the process of acting and this has led to other concepts, e.g. 'arena theatre' (theatre-in-the-round) with the audience completely surrounding a stage area, centrally located, and 'total theatre' with a number of acting areas, more or less completely surrounding the audience (at different levels), not to speak of 'wide stage theatre' where the acting may pass horizontally from one area to another.

It has been an object of study for many theatre architects and designers to create theatre spaces where such variations of stage v. audience configurations could be realised at will, preferably by mechanical devices which shifted from one theatre form to another.

In practice, of course, the number of different possibilities has to be limited and very often the main possibilities are two or three different forms, e.g. proscenium stage with different designs of forestage—'thrust stage' and 'open stage'. Thrust stage is akin to the Elizabethan theatre and has acquired popularity, especially in the English speaking world (for Shakespearean drama), whereas the open stage is a concept which brings stage and audience together into one space by sacrificing some of the scenic options of change that require a fly tower.

Incidentally, the open stage concept presumes a common ceiling for stage and audience areas and this has important consequences for the acoustical conditions of the theatre. The immense fly tower space, in most cases crammed with curtains, draperies etc. represents, acoustically, an almost 100% absorbing area situated immediately above the actors (singers), subtracting useful early sound energy from the resulting sound field. It is very seldom in more conventional theatre design that the architect (let alone the acoustician) has had any say in the design of the stage area and the stage machinery. This has usually been left completely to the mercy of 'stage people'—designers, consultants, technicians and contractors—with the well known result that every centimetre of the stage depth dimension has been occupied by drop lines, cycloramas, light bridges, etc.

Some attempt to break this monopoly of design has been made over the years, but so far with practically no success, probably because the projects have been largely of a traditional proscenium type. The innovations have mainly been concerned with forestage facilities, including movable splays, drop lines etc., but also in the forestage area the specific acoustical requirements have often been neglected in favour of the visual aspects.

As an example of a well defined *multiform* (or mainly dualform) theatre where the acoustical design was integrated with the theatrical (vision, lights etc.) the recently inaugurated Wintergarden Theatre in London is described. An exclusive *multipurpose hall* of recent design (with musical pretensions) is the congress/concert hall of Abidjan (Ivory Coast). As an example of a mixed multiform/multipurpose theatre the City Theatre of Malmø (Sweden) is of interest. Originally, the acoustical problems were not taken too seriously and the theatre is now being thoroughly investigated as the first step for a planned renovation. Finally, the project of a multiform/multipurpose theatre for Umeå Municipality (Sweden) must also be included to illustrate the elaborate integration of theatrical and acoustical design.

Wintergarden Theatre

The project for a new theatre at the site of the old Wintergarden Theatre in Drury Lane in the West End of London, started in 1964, was in fact quite sensational. A theatre designer of international reputation (the late Sean Kenny of London) had proposed the building of a new theatre without any fire curtain so that a choice between 'open stage' and 'arena stage' could be incorporated in the design.

To propose is easy, but to persuade the authorities (especially the fire authorities) to accept and to get private investors interested in an advanced scheme like this is extraordinary. Actually, it was the first theatre in London for 80 years or more, purpose built, having in its concept the idea of experiment as well as commercial use. It did take quite some time and trouble to get this theatre built but it was completed and it opened in the Autumn of 1972.

The capacity is more than 900 seats and the seating arrangement includes over 200 seats on a large revolve which may be swivelled around through 180° so that the stage part of the revolve occupies a central position in the auditorium (see Fig. 8.1).

The main problem, acoustically, apart from adjusting RT to a value appropriate for speech, centres on the design of the ceiling which acts as the important reflecting surface. Since the fly tower in this project was reduced and the use of fly lines limited, the intervening acoustical ceiling might be designed without secondary considerations, except lighting arrangements. Ideal reflecting surfaces, applicable to both the 'open stage' and the 'arena stage' situation, were considered to be a series of convex discs (or cylindrical

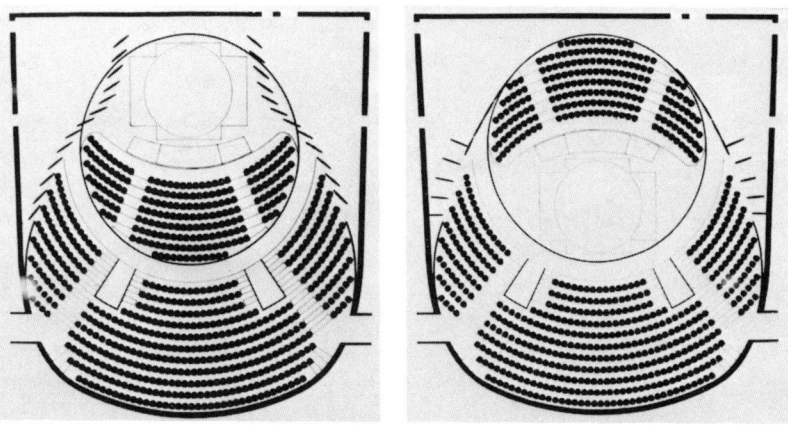

FIG. 8.1.   Wintergarden Theatre—plan, showing alternative arrangements.

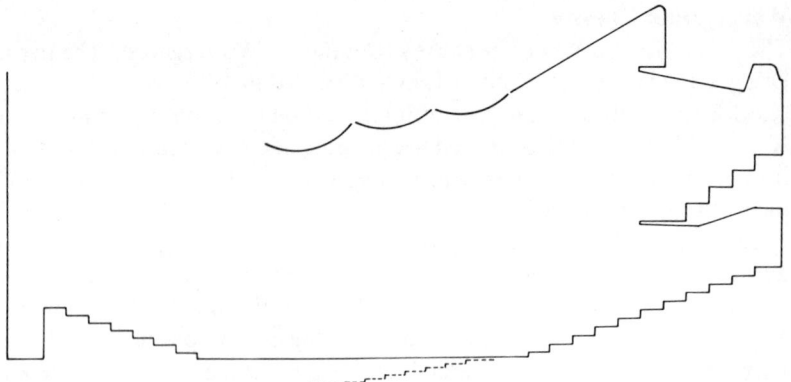

FIG. 8.2.   Wintergarden Theatre—longitudinal section, original concept.

reflectors) which would diffuse the sound in all (or both) main directions. This was the original concept (see Fig. 8.2). In practice one had to accept the fact that there should be several slots and openings for lights and pro-jectors; moreover, the cylindrical reflectors had to be subdivided for practical and aesthetic reasons into louvres of less width and of less pronounced convex shape.

These adaptations led to the final solution, as illustrated in Fig. 8.3.The louvres have a considerable freedom of orientation which in practice will be used more to accommodate lighting requirements and less to fit the acoustical requirements of sound reflection.

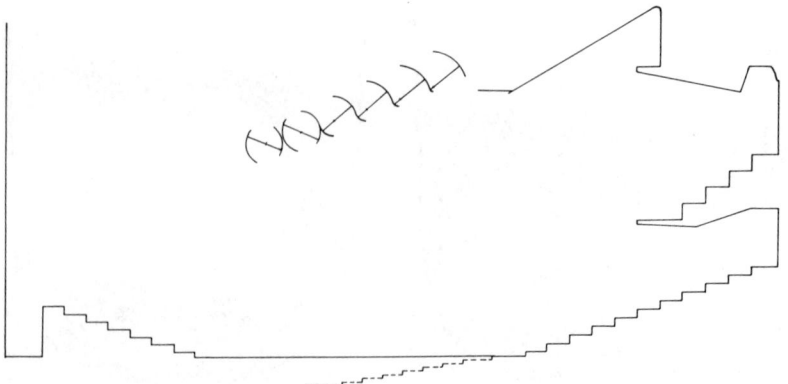

FIG. 8.3.   Wintergarden Theatre—longitudinal section, final solution.

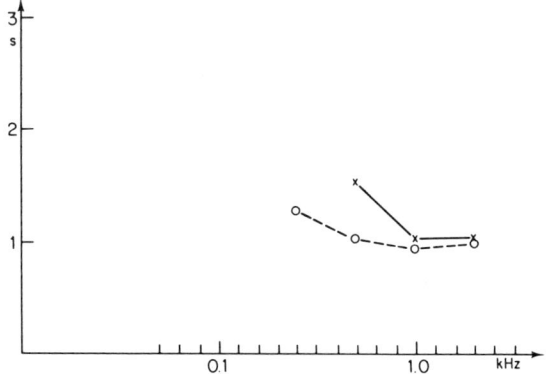

FIG. 8.4.   Wintergarden Theatre, RT (s), – – – and EDT (s) ——— v. frequency (kHz),
open stage.

The curvature of the auditorium would have resulted in pronounced focus effects if the rear wall had not been treated with strongly absorbing material which should be effective over the entire frequency range. This was accomplished by applying mineral wool spaced from the concrete walls and covering it with a heavy but open fabric as a surface treatment.

The completed (empty) auditorium has RT values in the range 1·2–0·9 s over the frequency range 250–4000 Hz (see Figs. 8.4 and 8.5). The values will change only slightly when the auditorium is occupied due to the heavily upholstered chairs. It was not feasible to make measurements of articulation but this property has subjectively been judged as satisfactory

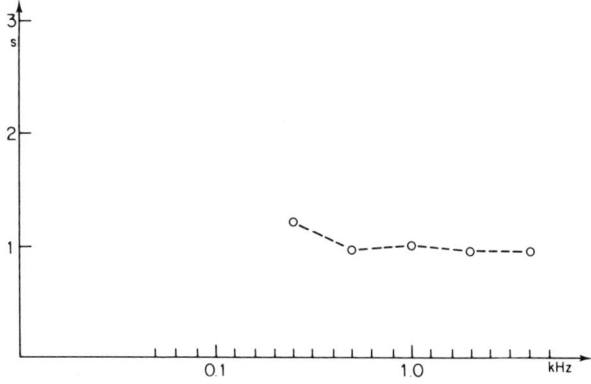

FIG. 8.5.   Wintergarden Theatre, RT (s) v. frequency (kHz), theatre-in-the-round.

over the entire audience area. The test of ambient noise level from air handling units did not give completely satisfactory results, particularly in the balcony area. Recommendations were given to correct the sound isolation of the fan units.

Palais de Congrès, Houphouët-Boigny

One of the former French dependencies which gained national independence, the Ivory Coast with its capital city, Abidjan, has already shown considerable development. The development was to include a National Building complex and should provide for congress, concert and theatre. An Israeli architect of high reputation (Heinz Fenchel of Tel Aviv) was chosen to be responsible for the design and was joined by a team of consultants from several European countries for the duration of the project (1970–73). The design was, to say the least, acoustically problematical. The main shape in plan (a spider's web?) and the exterior height of the building were more or less fixed by the architect's conception of the scheme; artistically admirable, but not very much concerned with acoustics. On the other hand, the management of the company involved in construction for the government was deeply concerned that the scope of the hall should not be confined to congress but that the musical acoustical quality should be emphasised and that future performances by visiting symphony orchestras of international repute should be envisaged.

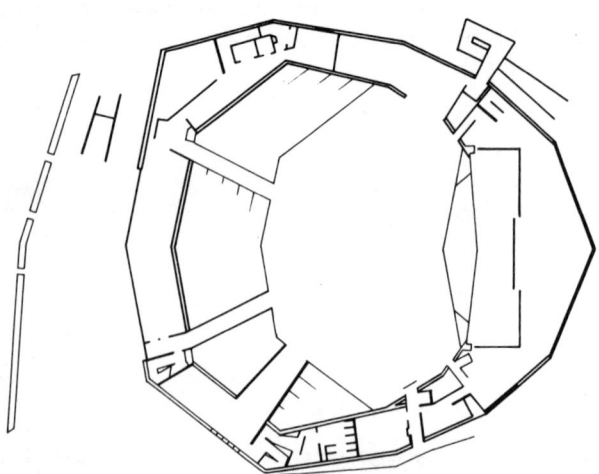

Fig. 8.6.   Palais de Congrès, Houphouët-Boigny—plan.

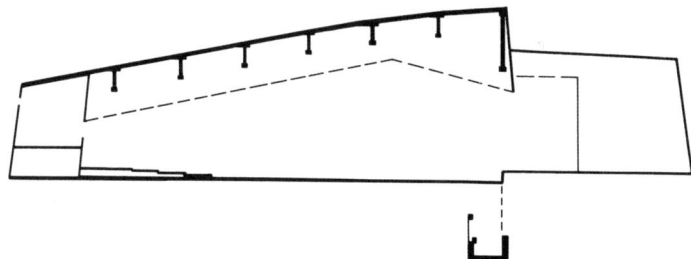

FIG. 8.7. Palais de Congrès, Houphouët-Boigny—longitudinal section.

Looking at the layout (Figs. 8.6, 8.7 and 8.8) it was obvious that the ratio of width to height was the fundamental problem. The interior had a low, conically shaped, suspended ceiling of maximum height 11 m (average height 8 m). These heights should be compared with an average figure of the width of approximately 40 m, which clearly indicates the seriousness of the situation. The only way of improving this situation was to consider the possibility of designing the interior ceiling as a decorative, but acoustically transparent, ceiling, and of integrating the volume above this ceiling with the main volume of the hall. Even if this were feasible the ratio of average height to average width would only improve to something like 14/40, which is still considerably less than the ratio found in most concert halls.

However, by reducing the distance between the upper part of the side walls (vertical surfaces) to approximately the width of the orchestra enclosure the ratio of height to width in the upper part of the hall would be improved to 14/26. For the benefit of the orchestral sound it was further recommended that reflecting surfaces should be installed around the stage and used whenever a musical performance was to take place.

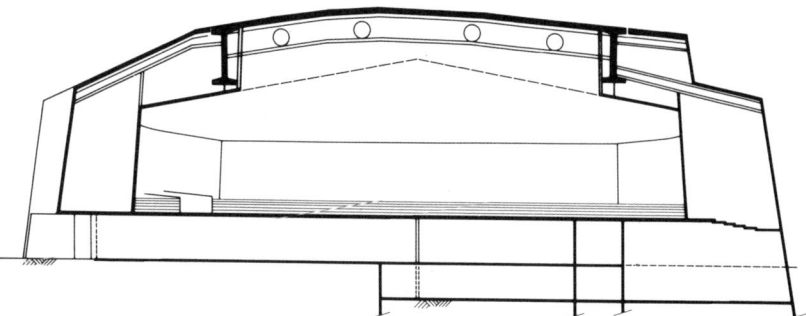

FIG. 8.8. Palais de Congrès, Houphouët-Boigny—cross section.

To study the acoustical problems of this hall it was recommended and accepted that a 1/10 scale model of the hall should be built and the criteria RT, EDT and II used to evaluate the acoustical quality.

The suspended, inner ceiling of the hall has a large number of solid beams, radiating from a centre piece (see view of the model, Fig. 8.9). It was recommended that between the beams an open grill ceiling of expanded metal be constructed and above that there would be catwalks and air ducts.

FIG. 8.9.   Palais de Congrès, Houphouët-Boigny—view of the model.

For the model ceiling only the solid beams of the suspended ceiling were simulated. The panelling on the vertical side walls (of the upper volume) which supplied low frequency absorption was also simulated in the model whereas catwalks and ducting were neglected.

Measurements in the model indicated quite clearly that the beams of the suspended ceiling had only marginal influence on the values of RT and EDT; EDT values showed only small variations with the location and most were slightly in excess of RT values . II values were in excess of 1.0. To check the influence of a *solid* ceiling, the model ceiling was changed by mounting hard fibre board on top of the beams. The subsequent testing showed values of EDT reduced considerably and also showed values of II considerably below 1.0. These results were interpreted as indicating that the transparent ceiling was a definite improvement.

Since the main purpose of the hall was for congresses and conventions it was important to design a high quality public address system combined with a simultaneous interpretation system.

Sound columns with highly directional distribution patterns (in the

FIG. 8.10.   Palais de Congrès, Houphouët-Boigny—view of the hall.

vertical direction) were located above the stage and close to the centrepiece of the suspended ceiling. The cluster of columns at the centrepiece were adequately delayed by applying a time delay system (with an acoustical rather than an electrical delay circuit, produced by Siemens, Karlsruhe).

The hall was inaugurated in the Autumn of 1973, but the inauguration programme did not include symphony concerts. Views of the hall are shown in Figs. 8.10 and 8.11.

Testing of RT, EDT and II showed that the target for RT (around 1·6 s) had not been reached. RT varied from 1·35 s (at 250 Hz) to 1·19 s (at 2000 Hz). RT v. frequency is shown in Fig. 8.12. The reason why RT in the

FIG. 8.11.   Palais de Congrès, Houphouët-Boigny—view of the hall.

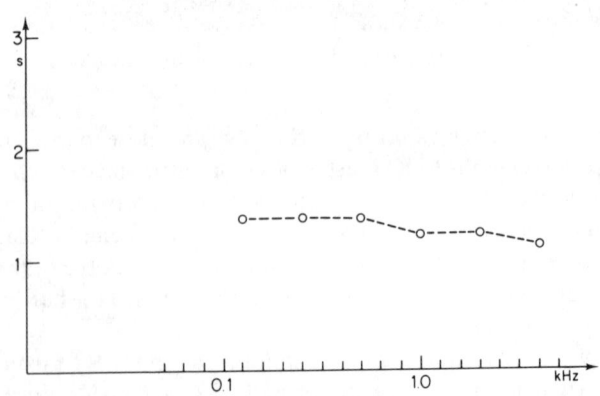

FIG. 8.12.   Palais de Congrès, Houphouët-Boigny—RT (s) v. frequency (kHz).

hall is lower than expected can only be guessed. It is believed that the absorbing effect of ductwork, catwalks and maybe also the expanded metal ceiling has been underestimated. Perhaps also, the wall panelling above the suspended ceiling has some high frequency absorption. The recommendation to spray all surfaces above the suspended ceiling with a glossy paint has not so far been carried out. Otherwise the testing indicated that values of EDT and RT are close to each other and that variations between different locations are small. Values of II are slightly above the values measured in the model and are above 1·0 for all frequency bands tested (250–2000 Hz).

Since the opening, the hall has been used frequently for conventions (with up to 1500 delegates) and has also been used for recitals and for entertainment, but a full concert with orchestra has not yet taken place.

Malmø City Theatre

The City Theatre of Malmø has many interesting features and it is worthwhile recounting how this design developed and what inspired it.

To understand this we shall go a little further back, to the beginning of this century and to an idea promoted by the French author Romain Rolland. He wanted to realise the idea of a 'People's Theatre', something of the same kind as the Greek Antique Theatre, where performances, People's Festivals and the works of the great classical authors would be performed before large audiences (maybe several thousands) seated in one huge, homogeneous audience area without intervening barriers (i.e., no loges, balconies etc.). The idea did not materialise in France but through the enthusiasm of the famous theatre producer Max Reinhardt, it was realised in Berlin in the form of 'Das grosse Schauspielhaus', with a capacity of 3000 seats. A large forestage, practically no proscenium frame and a very wide stage opening were the dominating features. The realisation of the grand idea was far from successful; the acoustics were nearly disastrous, since half the audience did not hear the actors, and although sight lines were excellent the contact between actors and audience suffered on account of the large distances. The house did not survive these problems and after a period of rebuilding, to improve the acoustics, it ended up as a variety theatre and, incidentally, was destroyed during the war.

But the dream of Romain Rolland and Max Reinhardt outlived the experiment and in the twenties and thirties many Russian theatres were built along these lines. Also, a young Swedish theatre consultant (Per Lindberg), working with a team of architects, designed a project for the city of Malmø with features taken from the Berlin Theatre House. Realising that it was important to limit the capacity they settled for a 1600 seat theatre and also introduced movable walls which permitted different sizes and shapes of the auditorium (reducing the number of seats to 1100, 600 or even 400; see Fig. 8.13) They retained the idea of a forestage protruding into the auditorium and also the idea of the very wide stage opening. They also had the well defined aim of making this theatre a musical (lyric) theatre by providing forestage elevators, which when taken down created space for an orchestra

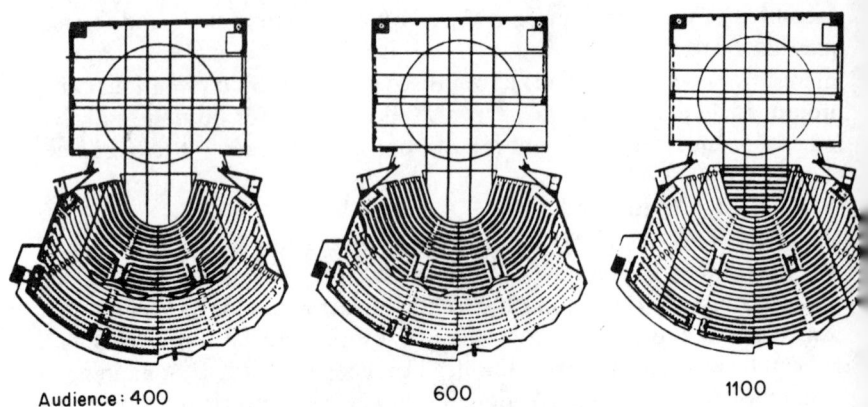

Audience: 400   600   1100

FIG. 8.13.   City Theatre of Malmø—plan, capacity audience various subdivisions of seating area.

pit and, finally, they designed an orchestra platform at stage level so that the auditorium could be used as a concert hall. This theatre was built during the war and was inaugurated in 1944, towards the end of the war.

With regard to the performance of drama, experience has shown that a capacity of 1000 is about the upper limit (as also experienced in other drama theatres). In the case of opera the experience is that although the stage, visually, is very well suited for grand opera, there are definite acoustical problems, not only for the singers, but also for the musicians and there is a problem of balance between singers and orchestra. In the performance of symphonic music there are problems of uneven distribution of sound and of lack of reverberance.

Recently, the management of Malmø City Theatre decided to let the theatre enter a phase of renovation and two items especially attracted attention; first, to modernise the lighting system and secondly, to improve the acoustics. As for the second improvement it has been emphasised that the main targets are to improve conditions for opera and for drama, whereas symphonic music should be of less concern (since the city of Malmø is planning to build a concert hall anyway).

It was considered imperative to make a thorough investigation of existing acoustical conditions in the hall and also to investigate a 1/10 scale model of the hall by the same methods. The proposed changes may then be implemented in the model and tested before they are finally adopted and carried out in the hall.

The fact finding, then, had to start with the present situation and to make a series of measurements in the auditorium as it is today. The important question of which criteria to apply has prompted the currently existing interest in the development of new criteria—also in connection with other projects—criteria which would possibly be more suited than EDT to indicate local variations within a hall. This will be discussed in the next chapter.

Umeå Theatre

A rapidly developing community in the province of Norrland, Sweden, is the city of Umeå. Fifteen years ago it also became the location of one of Sweden's new universities. The growing interest in several kinds of music, ballet, drama and opera, led to the foundation of a regional opera company, which in turn is looking for a permanent home.

This was the starting point for a project (not yet realised) which would have as its nucleus a theatre/concert hall with a capacity of 600–800 seats. Although involving a relatively small hall, it was a very ambitious programme in terms of the number of different possibilities the hall should provide, a genuine multiform/multipurpose situation. The list comprises: proscenium stage for opera, the same for drama, plus a regular concert hall. Further possibilities are: arena stage, Elizabethan stage and open stage. One of the main mechanical features to make this great flexibility possible is a modular system of floor elevators, occupying a considerable part of the total floor space and allowing a great variety of different floor levels to be combined. The design consultant for this theatre was Miklos Ölveczky of Malmø. Other mechanical devices include a movable proscenium frame, movable side reflectors at the proscenium and movable overhead reflectors in the stage area, which serve visual as well as acoustical purposes. Figures 8.14, 8.15 and 8.16 illustrate the theatre arrangements for different situations.

It is easy to see that this project may profit considerably by combining the design stage with an acoustical model investigation. The Umeå Theatre Committee, organised by the city municipality, agreed to this proposition

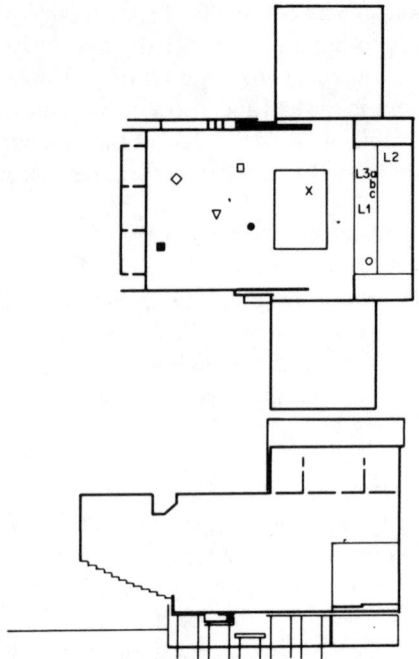

FIG. 8.14. Umeå Theatre—plan, and longitudinal section for Concert Hall situation. Sound source: L1, L2 and L3.

Microphone position: podium ○
M1 ×
M2 □
M3 ▽
M4 ◇
M5 ■
M6 ●

so that a programme for model testing could be set up. It was decided to concentrate on three of the most important applications which, in order of priority, are: 1, Opera, proscenium stage; 2, Drama, proscenium stage; and 3, Concert hall.

Realising, however, that the acoustical problems of the concert hall situation might outweigh the problems associated with the other main purposes, it was decided that the model investigation would start with the concert hall. The model would be made so that the switching between the three situations was easy to accomplish. A view of the model is shown in Fig. 8.17.

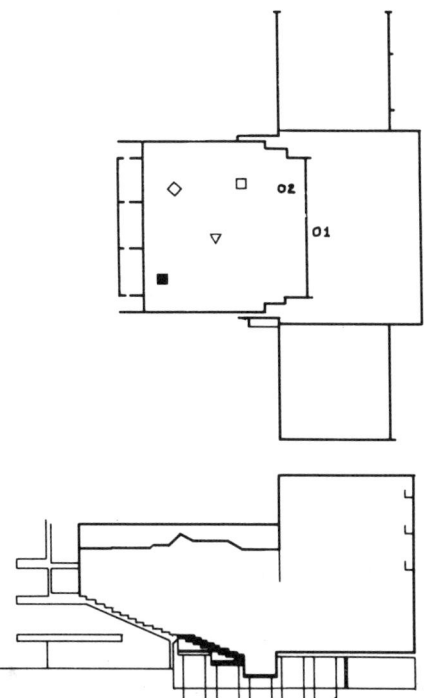

FIG. 8.15. Umeå Theatre—plan, and longitudinal section for Opera Theatre situation. Sound source: O1 and O2.

Microphone position: M2 □
M3 ▽
M4 ◇
M5 ■

First a decision had to be made as to which shape the concert hall should have. Since the volume was important in this case the height of the (basically) rectangular hall had to be considerable and it was thought appropriate to carry this height through into the stage house end so that the closing of the upper fly tower would be at the same level as the ceiling in the main body of the hall. Whether this closing off should be nearly 100 % or could be considerably less was a problem for the model testing to solve.

In the proscenium situation visual requirements demanded oblique surfaces at both sides of the stage opening. It was anticipated that they would serve a useful purpose as hard reflecting surfaces for the sound distribution in drama performances. In contrast, when using the

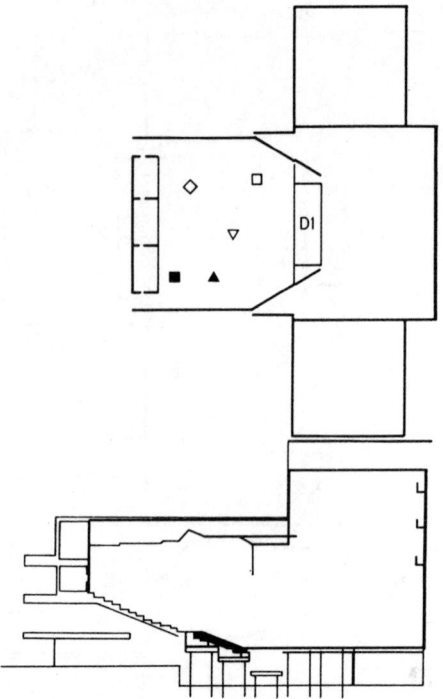

FIG. 8.16.   Umeå Theatre—plan, and longitudinal section for Drama Theatre situation. Sound source: D1.

Microphone position: M2 □
M3 ▽
M4 ◇
M5 ■
M8 ▲

proscenium arrangement for opera performances it was thought preferable to have the reflecting surfaces at both sides of the stage as stepped surfaces, parallel to the side walls of the hall.

However, visual requirements do not change from proscenium-drama to proscenium-opera so the discrepancy between acoustical and visual requirements would have to be solved by a combination of sound transparent and sound reflecting surfaces, all of which were to be moved away and stored at the side walls whenever there was a concert. Also the entire proscenium frame would in this case have to be moved.

Prior to the model testing, consideration had to be given to the choice of

FIG. 8.17. Umeå Theatre—view of the model.

criteria which should be adopted for the measurements. Evidently, this multipurpose situation of the Umeå Theatre which in principle resembled that of Malmø Theatre called for more than one criterion. The validity of the EDT criterion is well established but EDT is insufficient in cases where a correlation with articulation is desired.

Surveys of the development of objective criteria and their correlation with subjective assessment have been undertaken and previously reviewed on a couple of occasions. Now is the time to supplement those reviews with the latest information and to make a fresh approach, which is outlined in the following chapter.

BIBLIOGRAPHY

Subject	Reference
Theatre buildings and theatre forms	Ölveczky, M. In: *Teaterteknik*, Nordisk Kultursekretariat, Copenhagen, 1975.

DATA FOR WINTERGARDEN THEATRE, LONDON

Year of completion: 1972
Volume: $14\,000\,\text{m}^3$
Total area: $875\,\text{m}^2$
Number of seats:

Main tiers: 504
Revolve: 206
Balcony: 215
Total: 925

RT and EDT measured in empty theatre:

	250	500	1000	2000	4000	(Hz)
EDT (open stage)	—	1·5	1·0	1·0	—	(s)
RT (open stage)	1·25	1·0	0·9	0·95	—	(s)
RT (arena stage)	1·2	0·95	1·0	0·9	0·9	(s)

General Description:
  side walls (*stage and auditorium*): plaster
  rear wall (*auditorium*):   75 cm mineral wool covered with fabric
  ceiling (*auditorium*):   plaster on expanded metal
  ceiling (*stage*):   movable louvres
  floors: vinyl on concrete
  seats: upholstered

DATA FOR PALAIS DE CONGRES, HOUPHOUËT–BOIGNY (ABIDJAN)

Year of completion: 1973
Volume: $17\,000\,\text{m}^3$
Total area: $1700\,\text{m}^2$
Number of seats: 2000

RT and EDT measured in empty hall, octave values:

	125	250	500	1000	2000	4000	(Hz)
RT	1·35	1·35	1·35	1·2	1·2	1·1	(s)
EDT	—	1·3	1·4	1·2	1·2	—	(s)

General Description:

walls (auditorium):	wood panelling, some areas slotted
walls (orchestra stage):	lead vinyl curtains
ceiling:	open mesh
upper ceilings:	asbestos panelling, mineral wool
upper side walls:	asbestos panelling, mineral wool
floors:	parquet
stage size:	220 m²
seats:	upholstered (leather)

Chapter 9

Development of Criteria and Model Research, II

During the sixties there was growing concern about how to define acoustical criteria (objective and measurable) which could be correlated with subjective assessments of the acoustics of large halls.

We have already discussed and applied the criterion EDT, which belongs to a group that may be termed *decay criteria*. These include early decay time (0–10 dB), initial reverberation time (0–15 dB) and others which all have much smaller level intervals than the Sabine reverberation time (RT, 0–60 dB).

Another group of criteria is based upon the energy received during a certain interval in relation to the rest of the energy, or to the total energy of an impulse. These criteria are measured by emitting short impulses, e.g. from an electric spark source, and recording the impulse response at several locations in a hall (or in a model). This group may be classed as *energy criteria*. We have already mentioned the first criterion of the group: 'Deutlichkeit' (definition (Chapter 4)), which is the ratio of the energy delivered in the first 50 milliseconds to the total energy. This was introduced specifically as a criterion for speech articulation. Similar criteria for music were defined by Beranek and Schultz:

$$\text{Reverberant/early energy} \sim \text{`Hallmass'} \sim 10\log\frac{\text{energy}\,(50–\infty)\,\text{ms}}{\text{energy}\,(0–50)\,\text{ms}}\,(\text{dB})$$

and by Reichardt:

$$\text{`Hallabstand'} \sim 10\log\frac{\text{energy}\,(0–5)\,\text{ms}}{\text{energy}\,(5–\infty)\,\text{ms}}\,(\text{dB})$$

Another criterion related to both energy and decay criteria is the so called 'point of gravity' time, given by:

$$\text{'Point of gravity' time} \sim \frac{\int_0^\infty t p^2 \, dt}{\int_0^\infty p^2 \, dt} \text{(s)}$$

This criterion, introduced by Cremer and Kürer, weighs the contribution of individual 'energy elements' according to their time of arrival. It avoids the abrupt limits of the energy criteria which have no relation to the hearing process.

By extending the interval of the early energy from the first *fifty* milliseconds to the first *eighty* milliseconds the above mentioned criterion 'Deutlichkeit' becomes more suitable as a musical criterion (of musical articulation, so to speak):

$$\text{Clarity (C)} = 10 \log \frac{\text{energy} (0-80) \, \text{ms}}{\text{energy} (80-\infty) \, \text{ms}} \text{(dB)}$$

which has been introduced by Reichardt and his coworkers.

All these criteria are supposed to be measured with omnidirectional microphones so that reflections from all directions are considered equally valuable. However, during the sixties it became a well established belief that early, *lateral* reflections had a specific ability to create what has been called 'spatial impression', if they arrive within a time interval between *twenty-five* and *eighty* milliseconds after the emission of the impulse.

For the following discussion it is only necessary to introduce one criterion related to spatial impression, namely:

early *lateral* energy (25–80) ms/early total energy (0–80) ms

which is measured as a coefficient (not in dB).

When considering the abundance of different criteria and their apparent complexity you may envy Sabine and his simple criterion of reverberation time, and you may start asking what justifies this development of several criteria to characterise 'the acoustics' of a hall? However, the development in itself indicates that the subjective assessment of 'the acoustics' may depend upon several mutually independent qualities. The important question which has been thoroughly investigated in the last decade is in fact: how many different factors do we need to characterise 'the acoustics' of a hall, and is it possible to correlate these factors unambiguously with a corresponding number of objective criteria? This problem has been treated simultaneously by many research teams but the investigations have been especially extensive in three different research groups, located respectively in Dresden, Berlin and Göttingen. Many of their results are inconclusive but certain features are emerging which indicate how a group of criteria may be composed and how many individual criteria may be needed to characterise 'the acoustics' of a hall.

First, let us concentrate on the development which is characteristic of the Dresden group. Experiments with synthetic music samples, with direct sound plus differently delayed reflections and reverberation, made it possible (for Reichardt and Schmidt) to establish a scale of subjective evaluation of 'spatial effect' which could be correlated with an objective criterion called 'index of room impression'. The definition of this criterion is somewhat complex but in principle it is the ratio of all energy components contributing to the 'spatial sound' to all energy components contributing to the 'direct sound'. This ratio is expressed in terms of decibels.

When the subjective evaluation of 'spatial effect' was compared with the following four objective criteria:

(1)   'Index of room impression'
(2)   'Clarity'
(3)   'Point of gravity' time
(4)   'Hall| mass'

it became evident that the highest correlation coefficient was obtained with the 'index of room impression' (R). This has led the Dresden group to a definite acceptance of this index as *the* objective criterion for the quality 'spatial effect'. For quite a few years they have, however, been aware of the need for another criterion to correlate with the subjective quality of 'transparency' of music (actually divided into 'room transparency' and 'time transparency') and they have established that a high correlation exists between 'transparency' (in both senses) and the objective criterion of clarity (C). They have even established numerical values of C which may be considered as minimum values, in the sense that to secure transparency of music, the value of C should not be less than 0 dB (this specific value is stated as appropriate for music by Mozart, while other values are valid for other composers).

So far they have confined themselves to the firm establishment of the two criteria 'index of room impression' and 'clarity' without being specific about possible additional criteria to characterise other qualities of a hall.

At this point it seems better to postpone the account of the investigations in Berlin and Göttingen which will be taken up later, in Chapter 11.

The choice of clarity and index of room impression as *the* two deciding criteria has influenced the model testing at our laboratory for a couple of years and several models (plus a few halls) have tentatively been tested by measuring criteria of this kind. Clarity has been applied exactly according to the definition of the Dresden school whereas the 'index of room impression' has been modified and substituted with a simplified criterion based on the following reasoning.

It has been established that the two most important contributions to the subjective perception of 'spatial effect' are: 1, the contribution of lateral reflections in the time interval of 25 to 80 ms, and 2, the contribution of all reflections in the time interval of 80 to 160 ms. If you include these two contributions and express them as a fraction of the total energy (from 0 to 80 ms), this criterion will be highly correlated with the subjective evaluation of 'spatial effect'.

The definition of this simplified criterion, which we have called room response (RR), may be written:

$$RR \sim 10 \log \frac{\text{lateral energy}\,(25\text{–}80)\,\text{ms} + \text{total energy}\,(80\text{–}160)\,\text{ms}}{\text{total energy}\,(0\text{–}80)\,\text{ms}} \quad (\text{dB})$$

What exactly do we understand by the term 'lateral energy' and how are we going to measure this contribution? Reichardt and the Dresden group do not measure lateral energy. They measure *frontal* energy with a directional microphone. This energy, by definition, comprises all energy within a rotational angle of 40° (subtended by the source at the receiver). The simplest way of measuring lateral energy is by using a figure-of-eight microphone orientated so that it has its maximum sensitivity in the lateral directions, a method which has been adopted.

When collecting data from the measurement of C and RR it is important to be able to compare measured values with calculated values ('expectation values', calculated under the assumption that the decay process is perfectly exponential. The formulae to be used for calculating expectation values are given in Appendix V).

We shall now have to consider the implications for model testing of applying the criteria C and RR as a supplement to the testing of EDT.

We have already discussed the unavoidable limitations to the frequency range due to sound absorption in air at ultra sound frequencies. This absorption limits the frequency range for which we can expect to measure correct values of RT in a 1/10 model and we have restricted the frequency range to the octave bands: 1, 2, 4, 8 and 16 kHz. The highest of these bands (11–22 kHz, centred on 16 kHz) shows an air absorption which is no longer negligible and it is advisable to limit the frequency range further. With the measurement of RT and EDT it has become standard practice to use the octave 10 kHz (in a 1/10 model) as the upper limit and correspondingly to measure in the octave 1 kHz in full scale.

Some numerical calculations indicate that the influence of sound absorption in air is sufficiently limited for both EDT, C and RR if we limit the model testing to the 10 kHz octave. Another problem in model testing is the selection of microphones. By recording RT and EDT it is standard practice to use 6 mm or 12 mm condenser microphones, which could have been used if C only were to be measured. With measurements of RR where a figure-of-eight characteristic is required the search for a commercially available stereo condenser microphone shows that the smallest dimensions have a membrane diameter of approximately 3 cm. This is by no means ideal when considering the frequency spectrum of the model testing (octave band 7–14 kHz).

Also, the tandem arrangement of the two microphone units means that there will be diffraction phenomena around the capsules. For the figure-of-eight characteristic the polar diagrams are very similar at 1 and 10 kHz, whereas the omnidirectional characteristics differ considerably (since the linear dimensions of the capsules are comparable with the wavelength of sound at 10 kHz).

Another problem is the diffraction of the sound field close to the level of 'the heads of the audience' (in a model). This means that measurements must be taken with the microphones at a somewhat higher level. If the two capsules are orientated with the axis of the microphones vertical, diffraction phenomena will influence the two capsules to a different degree.

Measurements of C and RR also present some problems in full scale halls. Ideally they would have to be carried out in occupied halls but applying the pulse method in full scale (with a shotgun as sound source) is somewhat embarrassing (at least for the audience) although it has been done in a few cases. The development of a method whereby an orchestra is used as the sound source appears to be a possibility which may be realised— it has in fact been established for the decay parameters. However, the extended sound source represented by an orchestra in itself presents a problem when the aim is to measure criteria like C and RR since these criteria, strictly speaking, assume point sources. The variation in numerical value of the criterion C may prove to be quite considerable (and the same applies to the criterion RR), even with minor variations of the source location. It may be necessary either to resort to the application of a large number of microphone positions (and average) or to use a wider frequency spectrum (of two or even three octaves instead of one).

As an early example of trying out the test method of C and RR in a model, the Oslo Hall model should be mentioned. At this stage the frequency range applied was just limited by the A-weighted frequency curve. Values of C are distributed on both sides of the expectation value (C = $-2 \cdot 0$ dB) and they range from $-3 \cdot 3$ to $+1 \cdot 2$ dB. Values of RR are below the expectation value (RR = $-0 \cdot 3$ dB) and they range from $-3 \cdot 5$ to $-0 \cdot 4$ dB.

It was not feasible to make extended measurements of C and RR in the completed hall but on a single occasion a test was run in the empty but completed hall.

Values of C are definitely lower than the expectation value (the same as for the model) and they range from $-5 \cdot 5$ to $-3 \cdot 0$ dB. It has been found, however, on another occasion, that in the occupied hall, considerably higher values exist; it has not been possible to make a full series of measurements with an audience present. Values of RR are much higher than in the model (and much higher than the expectation value). They range from $+1 \cdot 3$ to $+3 \cdot 0$ dB. The problem with the discrepancy of RR values is apparently associated with at least two different sources of error. One is the diffraction phenomenon around the microphones in model testing. Another is the insufficient accuracy of calibration between the two channels of the tape recorder.

It was concluded that a considerable refinement of the measuring technique had to be undertaken and was essential for the actual application of the method, both in models and in halls.

BIBLIOGRAPHY

Subject	Reference
Reverberant/early energy	Schultz, T. J., Acoustics of the Concert Hall, *IEEE Spectrum*, June 1965, p. 56.
Hallabstand	Reichardt, W. and Schmidt, W., *Acustica*, **17**, 1966, p. 125.
'Point of gravity' time	Kürer, R., *Acustica*, **21**, 1969, p. 370.
Clarity	Reichardt, W., Abdel Alim, O. and Schmidt, W., *Acustica*, **32**, 1975, p. 126.
'Spatial Effect' (Raumeindruck)	Reichardt, W. and Schmidt, W., *Acustica*, **17**, 1966, p. 175.

Room Impression Index, R

Reichardt, W. and Lehmann, U., *Applied Acoustics*, **11**, 1978, p. 99.

Orchestra as sound source when measuring criteria

Schroeder, M. R., Gawron, H. J., Gottlob, D. and Möller, W. J., DAGA '76, VDI Verlag, Berlin-West, pp. 237–240.

Early lateral reflections

Schubert, P., *Hochfrequenztechnik u. Elektrotechnik*, **78**, 1969, pp. 230–245.

Barron, M., *J. Sound and Vib.*, **15**(4), pp. 475–494.

Chapter 10

Recent Projects of Varying Design

A factor common to a number of recent hall designs of average size (or less) is that they have all been tested with the aid of 1/10 scale models and also that the criteria applied have been extended from the previously applied parameters RT, EDT and II to the more recently adopted C and RR. Refinements of the measuring method have been introduced step by step, but only in one case has it been possible to arrange parallel testing between an already existing hall and its scaled model.

In spite of considerable efforts to establish conformity between values measured in hall and model the results have not been too promising and we shall have to consider the possible reasons for this lack of reproducibility.

The projects or halls to be mentioned (or further described) are: Concert Studio, Stockholm; Concert Hall, Dublin; Municipal Theatre, Umeå; and City Theatre, Malmø.

Concert Studio, Stockholm

The Swedish Radio for many years had its studios scattered all over the inner city of Stockholm and the Radio Symphony Orchestra had no concert studio of its own. When 'Sveriges Radio' finally built their big complex at 'Gärdet', just outside the inner city, they started with an office block, continued with a studio block and then switched to the television buildings —so the orchestra had to wait quite a while before the decision to create a concert studio was finally taken. Nearly 30 years after the war the design stage was reached and the studio was inaugurated in November of 1979. This studio will hold an audience close to 1300 so it will also function as a concert hall. Although the management was keen on maintaining that the prime purpose was radio transmission of symphonic music it was admitted that a

165

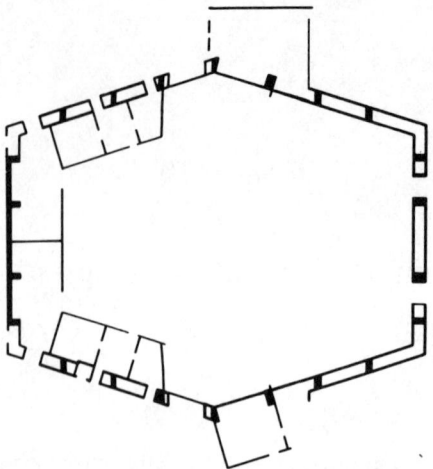

FIG. 10.1.   Concert Studio, Stockholm—plan.

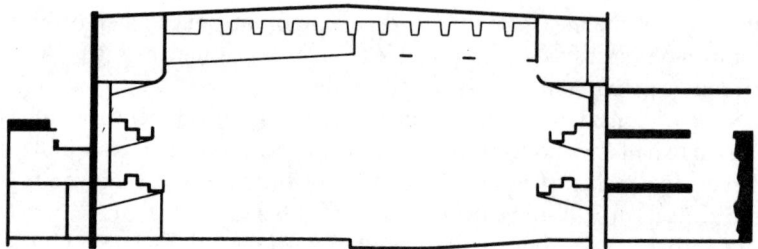

FIG. 10.2.   Concert Studio, Stockholm—longitudinal section.

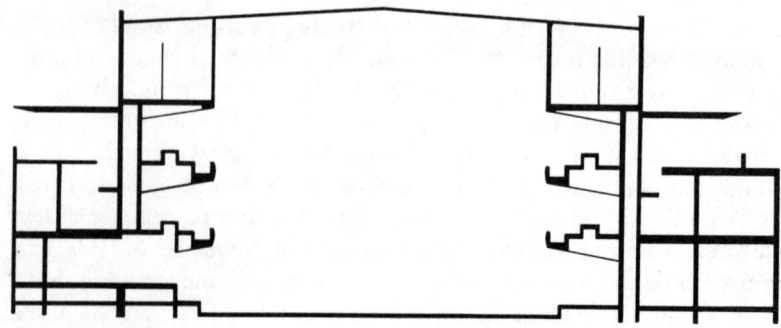

FIG. 10.3.   Concert Studio, Stockholm—cross section.

Fig. 10.4.   Concert Studio, Stockholm—view of model (1/100).

secondary purpose would be television transmissions of the orchestra. It was stressed, however, that the acoustical requirements for symphonic music should in no way be sacrificed for the functions of television.

The starting point for the design (Figs. 10.1, 10.2, 10.3, 10.4 and 10.5) was the shape in plan, and the classical rectangular shape had good reasons for being recommended. It was thought justifiable to change the rectangle,

Fig. 10.5.   Concert Studio, Stockholm—view of model (1/10).

however, and to modify it to a hexagonal shape which would provide side reflections in the rear part of the auditorium. (Incidentally, the adopted shape has some similarity to the gross shape of the Concert Hall in Sydney (see Chapter 6).)

The ceiling shape had to be diffusing and previously mentioned experiments undertaken in the model of the Oslo Hall (with 'beams across') influenced the ceiling shape of this studio. There were discussions on how the arrangement of lights and projectors over the podium area could be combined with an acoustically satisfactory diffusing ceiling and it was finally decided to make the ceiling above the podium lower and to allow a number of slots in this area so that lights and projectors could be mounted on bars to be lowered through the slots (whenever television events took place). The diffusion from the slots was considered an asset, acoustically.

The arrangement of the audience, on floor seating and two balconies along the whole of the perimeter, provides for intimacy. The seats behind the orchestra will not only be used as seating for the choir but also for the audience.

The model investigation in the 1/10 scale model comprised measurement of EDT and II, looking for possible echo effects and also for any undesired screening effect of the balcony soffits.

RT in the model was adjusted to a value slightly above 2 s (2·25 s) and the values of EDT in several testing positions at all levels were measured. It was found that between 95 and 100 % of these values were less than 10 % smaller than the value of RT. II was calculated from average values of EDT (in stage area and audience area) and numerical values were found to be a few percent below 1·0. Checking on possible echo effects revealed certain danger spots in the first balcony area close to the angling point of the side walls. This was particularly pronounced in the first version of the model where the sound control room (plus other service rooms) were located on both sides of this point and where the balcony seating was interrupted by large plane surfaces (Fig. 10.6).

This arrangement was subsequently changed and the control room was located at floor level close to the stage. The balcony and the seating could then be continued along the side wall so that the large plane surfaces disappeared. This reduced the echo tendencies (flutter) which were still traceable with the sound source in the centre front area of the podium. This became a question of making the wall panelling diffusing in the critical areas.

Examining the screening effect of the balcony soffits revealed, at certain positions, considerable changes in the picture of the impulse response when

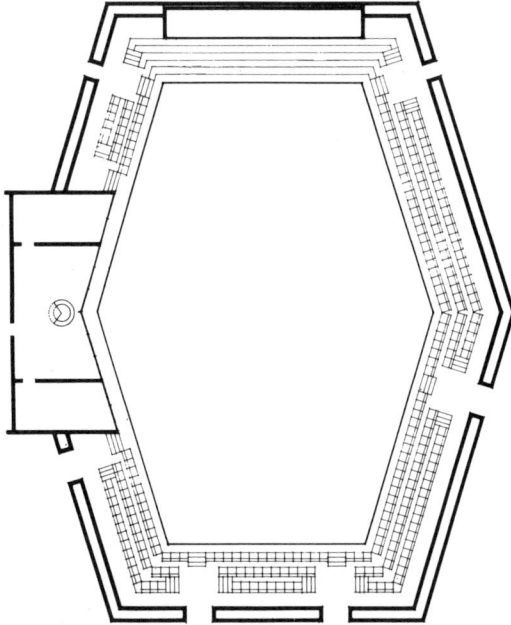

FIG. 10.6. Concert Studio, Stockholm—plan of lower balcony, first design.

switching the microphone position from the balcony front to the rear row. This led to a proposal which would reduce the overhang of the upper balcony and the change was also included in the final version of the project, and of the model.

In the auditorium itself it was decided to design the wall coverings as panelling of wooden strips (or planks, but with intervening slots). This design is similar to those in the concert halls of Sydney and Oslo and for the same reason—it allows minor adjustment of the RT v. frequency characteristic to be made after the completion of the hall and after the first acoustical and musical testing has been performed.

In the case of the Oslo hall the areas to be adjusted were located in places where first order reflections occurred. This meant that the changes in the absorption v. frequency characteristic also affected the frequency spectrum of these reflections. As already noted, the testing method for C and RR was not sufficiently refined at the time of testing in the Oslo hall so that changes of this kind could not be checked.

Tests of C and RR were also made in the model of the Stockholm Concert Studio and are graphically displayed in a C and RR diagram (Fig. 10.7). It

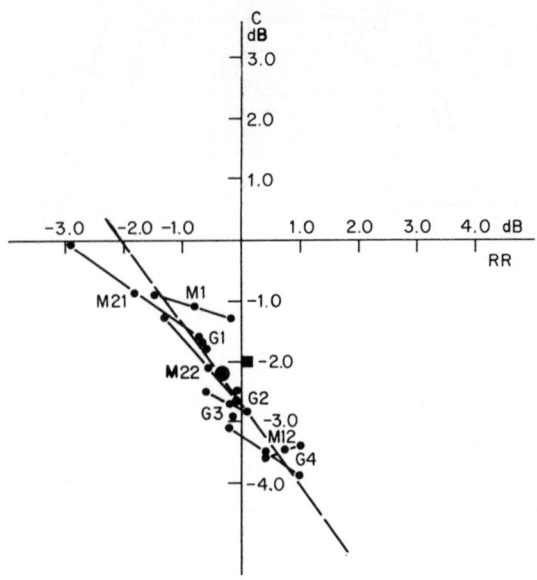

FIG. 10.7.   Concert Studio, Stockholm—C–RR diagram for model.
$C_{average}$ -2.2 dB, $RR_{average}$ -0.3 dB, ●;
$C_{expected}$ -2.0 dB, $RR_{expected}$ $+0.1$ dB, ■;
RT 0.225 s (dB(A) spectrum).

shows how two subsequent test series under identical conditions show random variations. The positions at floor level (G1–G4) obviously show less variation than the positions at the balconies (M11, M12, M21 and M22). It is also shown that the average values of the two series approach a regression line. The overall average values are close to the expectation values (calculated according to the formulae given in Appendix V).

The Dublin Concert Hall
This project has its origins in the idea of converting the old University Hall (Aula Magnum) into a modern concert hall. It was strongly recommended to the architect (Office of the Government Architect) to keep and restore, as much as possible, the original appearance of the ceiling, with the exposed beam system and their vaultlike transitions to the walls and the windows, and also of the upper walls (with rows of decorative pillars) since the hall would benefit acoustically from the diffusion from the surfaces. It was agreed that a new floor (sloping) and a balcony (completely surrounding the floor) should be added and adapted to the upper part of the hall. It was also

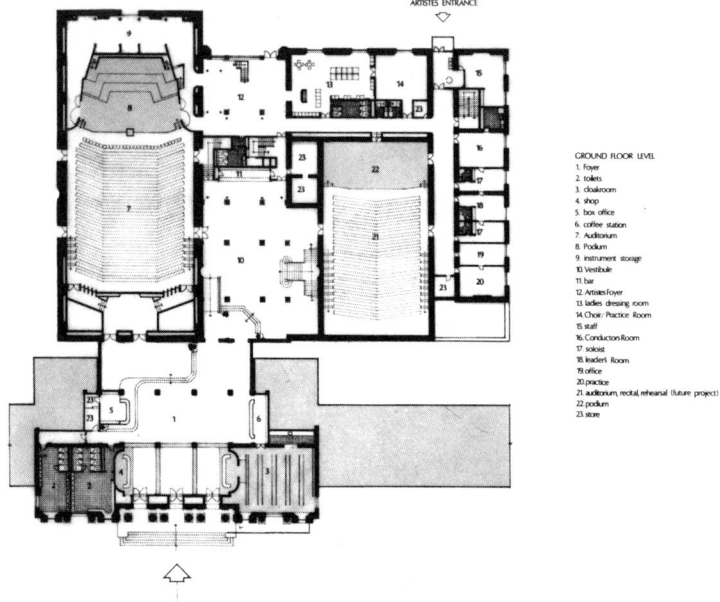

ARTISTES ENTRANCE

GROUND FLOOR LEVEL
1. Foyer
2. toilets
3. cloakroom
4. shop
5. box office
6. coffee station
7. Auditorium
8. Podium
9. instrument storage
10. Vestibule
11. bar
12. Artistes Foyer
13. ladies dressing room
14. Choir / Practice Room
15. staff
16. Conductors Room
17. soloist
18. leaders Room
19. office
20. practice
21. auditorium, recital, rehearsal (future project)
22. podium
23. store

MAIN ENTRANCE

FIG. 10.8.   Dublin Concert Hall—floor plan.

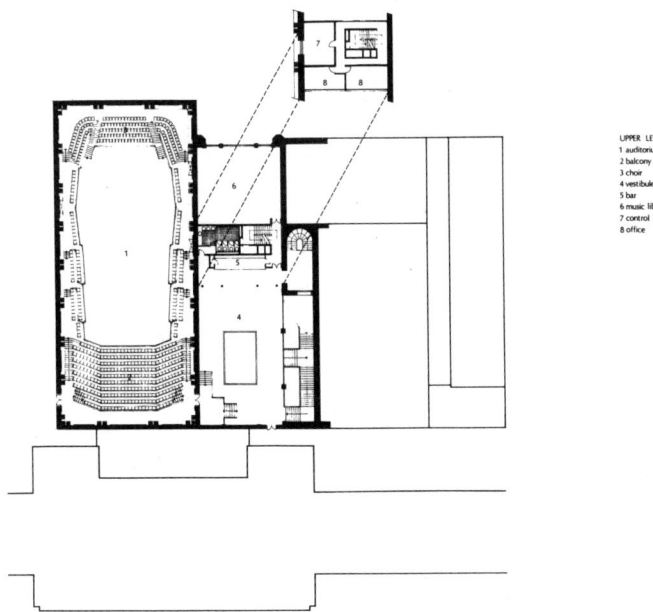

UPPER LEVELS
1. auditorium
2. balcony
3. choir
4. vestibule
5. bar
6. music library
7. control room
8. office

FIG. 10.9.   Dublin Concert Hall—balcony plan.

decided that there should be choir seats, and audience seating, behind the stage. A total seating close to 1300 seats is expected (see Figs. 10.8, 10.9, 10.10, 10.11, 10.12 and 10.13).

The testing in a 1/10 scale model aimed at establishing the variation of EDT values and also the values of C and RR for a number of positions. It was found that the variations of EDT were between -3.5% and $+9.5\%$ of

FIG. 10.10.   Dublin Concert Hall—view towards rear (sketch).

the average value of RT (0.20 s in 1 kHz octave). The II value is very close to 1.0 (0.97).

The values of C and RR are indicated in Fig. 10.14 where the average values and the expectation values are also given. The correlation between C and RR is apparently rather highly negative whereas the differences between average values and expectation values are a little higher than for the model of the Concert Studio, Stockholm. The tested sound spectrum has been limited to the 1 kHz octave.

FIG. 10.11.   Dublin Concert Hall—view towards stage (sketch).

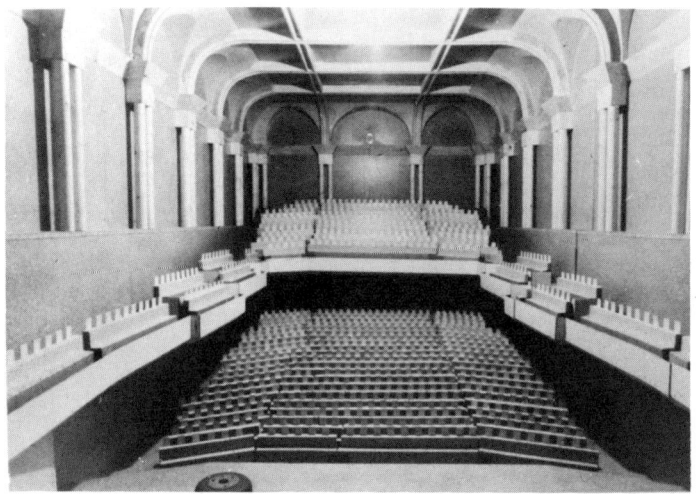

FIG. 10.12.   Dublin Concert Hall—model, view towards rear.

FIG. 10.13. Dublin Concert Hall—model, view towards stage.

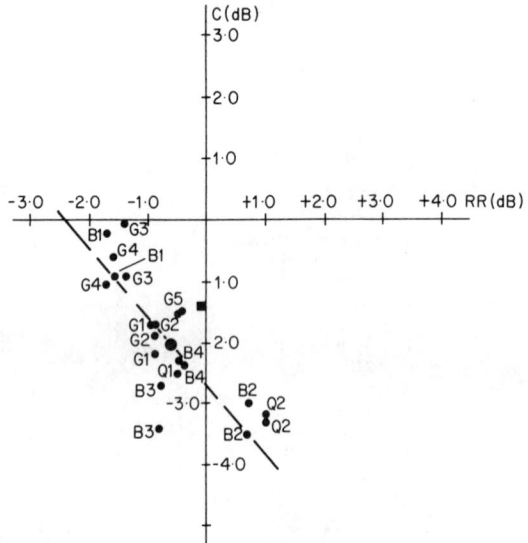

FIG. 10.14. Dublin Concert Hall—C–RR diagram, model.
$C_{average}$ $-2\cdot0$ dB, $RR_{average}$ $-0\cdot6$ dB, ●;
$C_{expected}$ $-1\cdot4$ dB, $RR_{expected}$ $-0\cdot1$ dB, ■;
RT $0\cdot2$ s (1 kHz octave band spectrum).

On the basis of the model testing it was recommended that the following changes should be made:

1.   To introduce a slope in the soffit ceilings at the underside of the balconies (side and rear),
2.   To modify the splays at both sides of the proscenium (in order to avoid sound pockets under the side balconies),
3.   To introduce some horizontal reflectors above the choir seating (in order to increase the clarity for chorus performances).

These changes were carried out and the values of C and RR were measured again.

While the sloping of the soffits at the side balconies and the changing of the splays at the sides of the podium both resulted in improvements of RR the slope of the soffit at the rear balcony did the opposite. These findings were reflected in the final recommendations.

The reflectors above the choir seating increased the clarity in the area below the reflectors but it was obvious that the area of the reflectors would have to be extended to cover the choir seating at the sides. This auditorium is now in the construction stage and is expected to open in 1981.

Municipal Theatre, Umeå

This project has been described in outline (Chapter 8) as an example of a modern multiform/multipurpose project. It was evident that model testing in this case could provide an interesting opportunity of investigating the variation of a number of criteria with the changing form of the interior. The three most important arrangements to be tested were: 1, concert hall; 2, opera theatre, and 3, drama theatre.

The main criteria to be applied for the investigation were EDT, C and RR. The actual layout in the model for the three main purposes has already been explained (see Chapter 8 and Figs. 8.15, 8.16, 8.17 and 8.18). The sketches also show the microphone positions used in the model testing.

In the concert hall situation the fly tower will be (partly or completely) screened off from the hall which will have seating also in the stage area. The tests, of EDT and also of C and RR, indicated that a 50% screening off would be sufficient and maybe even preferable to a complete closing of the fly tower. In the audience area a specific design of the ceiling was considered; an open mesh with a system of suspended vertical baffles in the ceiling space. This was considered because of the useful diffusing effect of such baffles (for example in the Ruben Dario Theatre of Nicaragua, see

Chapter 5). The measurements of C and RR indicated smaller overall variations with this ceiling design.

As an example of the many series of tests of C and RR in the concert hall arrangement one of the C–RR diagrams is shown (Fig. 10.15). The values indicated with crosses are averaged for two sound source positions on the stage, L1 and L2, whereas the values indicated with circles are averaged for

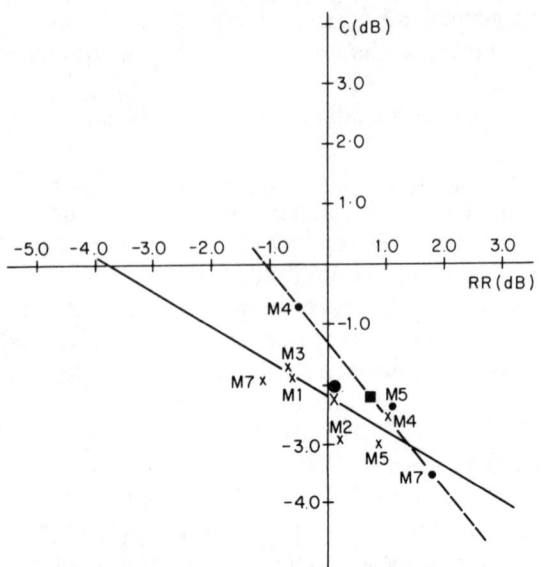

FIG. 10.15.   Municipal Theatre, Umeå, Concert Hall—C–RR diagram, model. Sound source on podium, closed ceiling.

L1 and L2 $C_{average}$ −2·3 dB; $RR_{average}$ +0·1 dB, × ;
50 % closed ceiling |
L3, a, b, c $C_{average}$ −2·2 dB; $RR_{average}$ +0·8 dB, ■;
                 $C_{expected}$ −2·0 dB; $RR_{expected}$ +0·1 dB, ●;
                 EDT: 2·25 s.

three sound source positions L3a, L3b and L3c likewise on the stage but only a few centimetres apart. Considerable variations occur with even small variations of the sound source (or the microphone) and this problem seems to be inherent in the testing method. Actually, with a spectrum of only one octave band width it seems impossible to escape such variations. This point will be further discussed in the next chapter.

In the opera theatre arrangement a regular proscenium stage (with fly tower) was simulated in the model. The proscenium reflectors (of which the angle can be varied) are kept parallel to the side walls of the auditorium in this case. The C and RR values were measured with the sound source located at the stage or in the pit. The C–RR diagram corresponding to these conditions is shown in Fig. 10.16. In accordance with the source position on the stage (01) or in the pit (02) we notice variations in C and RR values with

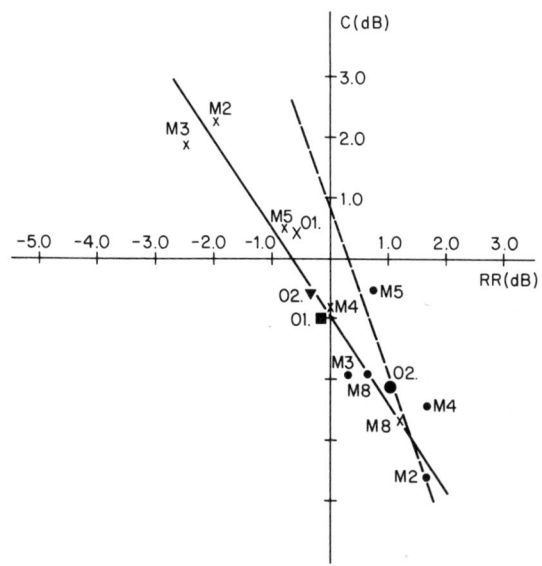

Fig. 10.16. Municipal Theatre, Umeå, Opera Theatre—C–RR diagram, model.
Sound source at stage:

$$C_{average} +0·4 \, dB, \ RR_{average} \ -0·6 \, dB, \ 01;$$
$$C_{expected} \ -1·0 \, dB, \ RR_{expected} \ -0·2 \, dB;$$
EDT: 1·88 s.

Sound source in pit:

$$C_{average} \ -2·1 \, dB, \ RR_{average} \ +1·0 \, dB, \ 02;$$
$$C_{expected} \ -0·6 \, dB, \ RR_{expected} \ -0·3 \, dB;$$
EDT: 1·74 s.

preferable conditions for opera (high clarity for singers, large room response for orchestra).

In the drama theatre arrangement the proscenium reflectors are angled to provide more directional sound from the stage to the audience. The corresponding C–RR diagram is shown in Fig. 10.17. We notice an average increase in the value of clarity which may be ascribed to the angling of the reflectors.

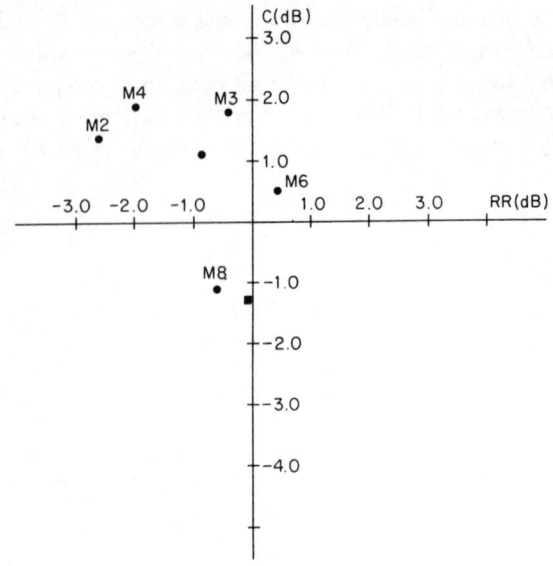

FIG. 10.17.   Municipal Theatre, Umeå, Drama Theatre—C–RR diagram, model.
Sound source at stage:

$C_{average}$ 1·1 dB, $RR_{average}$ −0·9 dB, ●;
$C_{expected}$ −1·3 dB, $RR_{expected}$ −0·1 dB, ■;
EDT: 1·97 s.

The Municipal Theatre of Umeå is still only a project. The prospect of building and completion is open for decision which may result in an opening date of 1982.

Malmö City Theatre

This theatre which is actually a multiform/multipurpose hall has already been described (Chapter 8) and it has been mentioned that acoustical problems have been experienced over the years since the opening in 1945.

The decision to update the theatre has now been taken, including also the decision to have a period of investigation of the acoustical problems. It has been stressed that the use of the auditorium for concerts is of lesser importance (since a concert hall will be built in Malmö, anyway), whereas the use for opera (operetta and musicals) is the main concern.

The present plans have also been indicated (in Figs. 8.13 and 8.14). The main problem is the large hall (A-hall) but the next smaller hall (B-hall) has to be considered. The A-hall has now been built as a 1/10 scale model which permits model measurements to be directly compared with full scale measurements. The criteria applied in the investigation are EDT, C and RR. Recently, it has been found useful also to include a specific criterion which relates early lateral energy to early total energy and which we shall denote as *Lateral Efficiency* (LE), defined as: LE ~ early lateral energy (25–80 ms)/early total energy (0–80 ms), which is measured as a coefficient (not in dB, see p. 159).

As usual, the problem when working with a scaled model is to ensure the acoustical similarity with the auditorium at least within a certain frequency' range. This is achieved by adjusting RT of the model so that the value after adjustment is close to one tenth of the value in the hall. Measurements of RT in the hall and in the adjusted model are shown in Fig. 10.18.

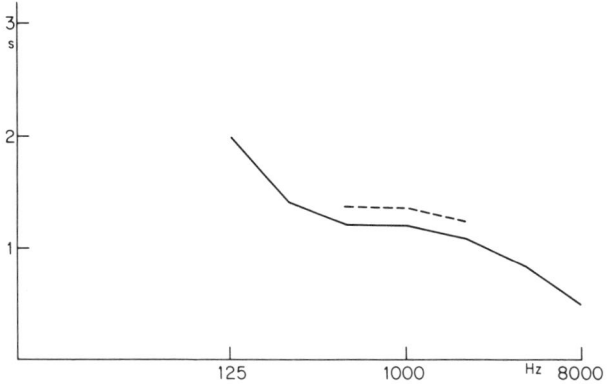

FIG. 10.18. City Theatre, Malmø—RT (s) v. frequency (Hz), auditorium (——) and model (---).

It is noted that in the 1 kHz octave (10 kHz octave in the model) the average values in the model and in the hall are close to the 1/10 relationship (model: 0·13 s, hall: 1·2 s). This has been considered to be close enough for comparative testing of the criteria. The sound source (pistol shot in the hall, electric spark in the model) was located either near the centre of the pit (L1) or near the centre of the stage (L3) while the microphones (AKG stereo in the hall, Schoeps stereo in the model) were moved to several positions (M1–M7, location diagram Fig. 10.19). A number of comparative tests in the hall and model indicated that while the values of EDT were reasonably close to conformity the values of C and RR did not follow any recognisable pattern when switching from hall to model. A number of reasons may be suggested for this lack of conformity. It has already been mentioned that

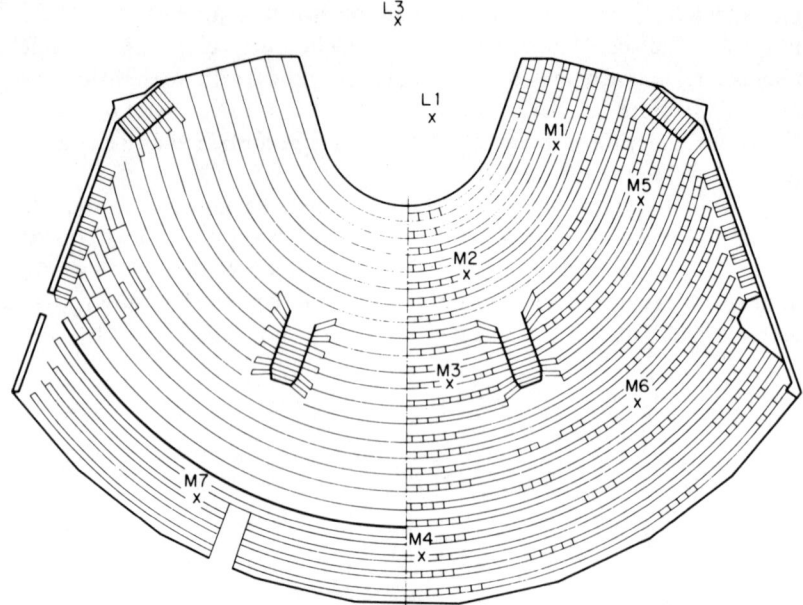

FIG. 10.19.   City Theatre—microphone location diagram.

the testing method (for C and RR) may in itself result in large random variations in the parameters even with only small variations in the position of the source (or microphone). One way of overcoming this problem is, of course, to take several test runs with small variations of source position. This is a cumbersome process which it has only been possible to use in a few cases.

Examples of C–RR diagrams obtained by this method for the hall and the model are shown (in Figs. 10.20 and 10.21), for the four microphone positions M2, M3, M5 and M7. Leaving the balcony position M7 out of consideration for the moment and taking (graphically obtained) average values for the three positions, M2, M3 and M5, and for the two source positions L1 and L3 (pit and stage) it is seen (on the graph for the model test) that the stage position gives much smaller values of the clarity and much higher values of room response than the pit position. The same tendency prevails for the auditorium test although the distinction is not so pronounced. The result in itself is peculiar because one would not expect such conditions.

Returning, for instance, to the corresponding diagram of the opera situation of the Umeå Theatre (Fig. 10.16) one finds more normal conditions, i.e., that the source position at the stage gives large values of C (small values of RR) whereas the opposite is the case with the source position in the pit.

The difficulties encountered by the testing of C and RR may, however, be more or less overcome in another way. If the test spectrum is enlarged from

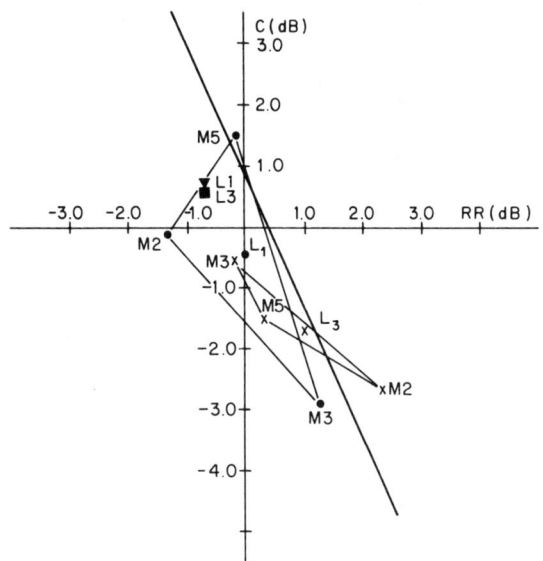

FIG. 10.20.   City Theatre—C–RR diagram, A-hall, Opera.
Sound source at stage   L3 ×, $C_{average(2,3,5)}$ $-1\cdot7$ dB
$RR_{average(2,3,5)}$ $+1\cdot0$ dB
Sound source in pit   L1 ●, $C_{average(2,3,5)}$ $-0\cdot4$ dB
$RR_{average(2,3,5)}$ $0\cdot0$ dB
Stage: $C_{expected}$ $+0\cdot6$ dB, $RR_{expected}$ $-0\cdot7$ dB, L3 ■
Pit: $C_{expected}$ $+0\cdot6$ dB, $RR_{expected}$ $-0\cdot7$ dB, L1 ▼
EDT: $1\cdot4$ s.

one to two (or even three) octaves it should be possible to avoid interferences and to establish more reliable values of the criteria. This, however, complicates the acoustical adaptation of the model.

Incidentally, working with two criteria like C and RR *simultaneously*, also adds to the difficulties and it was decided that some test series should be carried out with only one criterion. For these series it was thought most expedient to apply the criterion of *lateral efficiency* (LE).

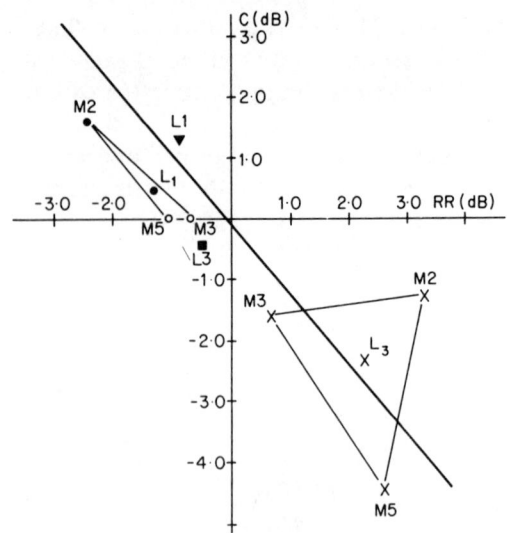

FIG. 10.21.   City Theatre—C–RR diagram, A-hall, Opera, model.
Sound source at stage   L3 ×, $C_{average(2,3,5)}$ −2·4 dB
                                        $RR_{average(2,3,5)}$ +2·3 dB
Sound source in pit   L1 ●, $C_{average(2,3,5)}$ +0·5 dB
                                        $RR_{average(2,3,5)}$ −1·6 dB
Stage: $C_{expected}$ −0·4 dB, $RR_{expected}$ −0·5 dB, L3 ■, EDT: 1·6 s
Pit: $C_{expected}$ +1·2 dB, $RR_{expected}$ −0·9 dB, L1 ▼, EDT: 1·3 s.

Since LE is important for both C and RR it is evident that any auditorium
(and any location within an auditorium) which exhibits large values of LE
must be advantageous and therefore it does seem quite natural to choose
LE as a single criterion. When resuming the discussion on criteria in the
next chapter (Chapter 11) we shall elaborate on this point.

Measurements of LE have been made both in the hall and in the model
and the results have been compared in a diagram (Fig. 10.22). Both source
positions have been applied. For some microphone positions this results in
large variations of the corresponding LE values. The spread, however, is
evidently around the conformity line, and this becomes quite clear for the
average values of pit and stage position of the source.

The average values for all five microphone positions (taken separately for
the pit and the stage position of the source) are indicated on the diagram
together with the expectation value of LE. It is noted that these average
values are considerably below the expectation value and this fact in itself
might be an indication of inadequate acoustical conditions.

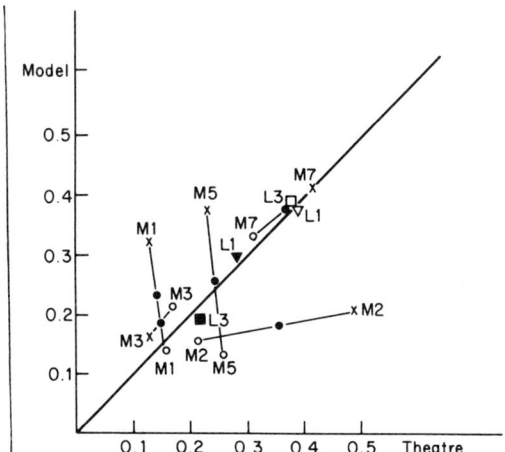

FIG. 10.22. City Theatre—comparison between LE values of Opera and model.
Sound source on stage, L3 ○
Sound source in pit, L1 ×
Average of pit and stage values: ●

	model	theatre	symbol
LE$_{average, L3 \ values}$	0·197	0·214	■
LE$_{average, L1 \ values}$	0·298	0·280	▼
LE$_{expected, L3 \ values}$	0·389	0·374	□
LE$_{expected, L1 \ values}$	0·377	0·383	▽

Microphone positions as in Fig. 10.19.

Similar measurements of LE have been made in some of the current
project models and the average values have been compared with the
expectation values. Values from the models of Dublin Concert Hall and of
Umeå Theatre (opera) are given in Table 10.1.

It is seen from the figures that in all four cases quoted the average values
are *larger* than the expectation values. When we look again at the diagram
of the Malmø Hall and compare it with its model, indicating LE values, we
notice that the only position where the value of LE comes close to the
expectation value is the rear balcony position, M7. No doubt, this fact is
associated with the particular shape of this auditorium (very wide and very
open fan shape) which does not favour side reflections; rather, the opposite.
All side reflections are more or less concentrated in the rear balcony area.
Looking at the definition of LE we see that the total sound energy (0–80 ms)
is the denominator so that a reduction of this factor is favourable. This is an

TABLE 10.1

COMPARISON OF MEASURED AND EXPECTED LE VALUES FOR TWO
THEATRE MODELS

	Average source position		Expectation source position	
	L1	*L2*	*L1*	*L2*
Dublin Concert Hall, model	0·405	0·480	0·396	0·396
Umeå Theatre (Opera),	O1	O2	O1	O2
model	0·418	0·468	0·393	0·391

indication that an increase in ceiling height, especially in the front part of the hall, may improve the conditions, and the values of LE.

The model research of this development project continues and further emphasis will be put on the evaluation of the criterion LE.

BIBLIOGRAPHY

Subject

Reference

Measurement of short time criteria, e.g. clarity, etc.

Reichardt, W. Private communication, 1978.

DATA FOR CONCERT STUDIO, STOCKHOLM

Year of completion: 1979
Volume: $12\,400\,m^3$
Total area: $750\,m^2$
Number of seats
    Orchestra level: 531
    1st balcony: 452
    2nd balcony: 382

    Total: 1365

RT (calculated):   1·8–2·0 (s)
RT (model test):   2·0–2·1 (s)
EDT (model test):   2·0–2·1 (s)

General Description:

walls:	wood panels, some areas with slots
ceiling:	gypsum board
floors:	parquet
seats:	upholstered
platform size:	210 m²

DATA FOR CONCERT HALL, DUBLIN

Expected year of completion: 1980
Volume: 13 000 m³
Total area: 830 m²
Number of seats
  orchestra level: 705
  balcony level: 580

     Total: 1285

RT (calculated)   2·0–2·2 (s)
RT (model test): 2·0–2·1 (s)
EDT (model test): 2·0–2·1 (s)

General Description:

walls:	wood panels, some areas with slots
ceiling:	gypsum plaster (on lath)
floors:	undecided
seats:	upholstered
platform size:	160 m²

DATA FOR MALMØ CITY THEATRE

Year of completion: 1945, expected renovation

Concert Hall and Opera Theatre (A-hall)
Volume: 9500 m³
Total area: 1060 m²
Number of seats: 1600

RT, EDT (measured in empty hall, 1/3 octave values):

	125	250	500	1 000	2 000	4 000	8 000	(Hz)
RT	2·0	1·40	1·20	1·20	1·10	0·85	0·50	(s)
EDT	—	—	1·40	1·30	1·05	0·80	—	(s)

Drama Theatre (*B-hall*)
Volume: 7400 m³
Total area: 800 m²
Number of seats: 1200

RT, EDT (measured in empty hall, 1/3 octave values):

	125	250	500	1 000	2 000	4 000	8 000	(Hz)
RT	—	1·10	1·05	1·10	1·20	0·90	0·60	(s)
EDT	—	—	1·25	1·20	1·15	1·00	—	(s)

General Description:

walls:	wood panels (rear wall corrugated)
balcony soffit:	perforated panels
ceiling:	wood panels on plaster ceiling
seats:	upholstered
pit size:	88 m²
stage size:	160 m²

Chapter 11

Development of Criteria and Model Research, III

We have already discussed some of the main conclusions of the Dresden research group on criteria. They may briefly be summarised as establishing two physical criteria corresponding to two subjective factors of judgement: 'Clarity' (C) corresponds to 'transparency'; 'Room impression index' (R) corresponds to the subjective judgement of room impression (or 'spatial effect').

Investigations of correlation between factors of subjective judgement and objective criteria have proceeded along different lines in Göttingen and Berlin.

One of the first main points to be studied in Göttingen (Siebrasse, 1973) was the possible number of individual factors inherent in subjective assessments and it was concluded that at least four different factors had to be considered. The first factor expressing 'consensus of preference' covered nearly 50 % of the total variance. With the choice of the second factor it was noticed that a certain division into groups of different taste became apparent.

On the basis of this purely psychological investigation the next step was an attempt to establish correlation between subjective factors of assessment and objectively measured criteria (Gottlob, 1973). Some of his (early) conclusions were that two criteria had high subjective relevance, namely early energy/remainder of energy (i.e. definition or clarity) and early nonlateral energy/early total energy. The latter may of course be replaced by the coefficient we have proposed, LE, since:

$$LE = 1 - (\text{early nonlateral energy/early total energy})$$

In later research (1975) the Göttingen group attempted to find out whether it is possible to distinguish between 'Halligkeit' (reverberance) and 'Räumlichkeit' ('spatial effect'). They concluded that it may not be possible at all. Therefore as a next step, they tried to establish a criterion which includes both the early lateral energy and the later, reverberant, energy. Actually, when we, following Reichardt, established a simplified 'index of room impression' defined as room response (RR) we based it on the findings of the Göttingen group, i.e. that such a criterion was highly correlated with a subjective scale of preference.

Meanwhile, the Berlin group (Lehmann, 1972) had applied another method of analysis (Wilkens, 1975) based upon a semantic scale of 19 different pairs of subjective judgements (e.g. 'hallig–trocken', 'gross–klein', etc.), which the judges had to apply when assessing musical samples.

The main conclusions of the Berlin group were that the three most important factors were 1, level of sound; 2, definition; and 3, tonal balance.

The Göttingen group, in their experiments, had always used a fixed level when presenting their musical samples and had considered the inclusion of the signal level as more or less self evident. The second factor 'definition' corresponds to the conclusion of Gottlob (early energy/rest of energy) but the third factor 'tonal balance' did not appear explicitly in the conclusions of the Göttingen group.

More recently, a member of the Göttingen group has presented the results of further investigations (Eysholdt, 1976) and this contribution seems to give some important clues. From the tests it appears to be proven that the value of RT is even less important than Gottlob assumed. In fact, it looks as if the limits of RT comprise an interval between approximately 1·4 and 2·8 s. Within these limits the value of RT is unimportant(!). It also looks as if the criterion EDT is definitely more critical, but that the range of acceptable values still comprises an interval of approximately 1·8 to 2·6 s. Within this interval subjective preferences are decided by other criteria, which depend either upon more limited time criteria (such as C, RR and LE) or upon totally different factors such as *level of sound* and *spectral balance*.

By varying the level of presentation it was established that a group of judges (totalling 14 listeners) was quite clearly divided into two smaller groups (of 9 and 5 listeners respectively) when assessing the optimal level. The individual preferences had a total variance of 9 dB but could be subdivided between one group with an (average) optimum preference of 84 dB(A), corresponding to a *mezzoforte* score, and another group with an (average) optimum preference of 88 dB(A) for the same score. This could only be taken as an indication of two different trends in individual taste and cannot be regarded as something which will affect the design of concert halls, generally, as an objective criterion. Out of 25 samples of concert halls compared for level difference, 19 halls had levels within an interval of only 3 dB(A), when they were subjected to the same radiated effect. The other, totally different, factor is the spectral balance. Spectral balance, of course, is affected not only by the individual hall, but by the musical score, by the composition of the orchestra (of the various instrumental groups) and also by the preferences of the individual conductor. Leaving aside the last three of these factors and concentrating on the objective influence of the hall, we considered earlier in this book the problem of spectral balance when we were discussing the RT v. frequency dependence, which occupied much of the time spent on investigations at Radiohuset, Copenhagen (Chapter 1) and which has also been mentioned in connection with other halls and theatres (e.g. new Metropolitan Opera, Chapter 5).

Having realised, however, the utter insufficiency of RT as an acoustical criterion for *large* halls in general it is necessary to look for some of the short time criteria which may be more applicable to spectral balance.

The investigation (by Eysholdt) mentioned above also treated the problem of spectral balance. It suggests a parameter which is fairly easy to vary in a steady manner and not too complicated to express in a simple formula.

The definition of *spectral density* (S) is 10 log of a quotient, which relates the A-weighted power spectrum of the upper frequency range (500–5000 Hz) of a musical sample (composed of 80 ms intervals of short time spectra) to the A-weighted power spectrum of the lower frequency range (50–500 Hz), measured in dB. According to this definition, the parameter S expresses the balance of energy between the upper and the lower frequency range and the value $S = 0$ means that the two ranges are exactly balanced. When a group of listeners was subjected to a number of musical samples for which the parameter S was varied it became apparent that their judgements of the tonal quality were not unanimous. Similarly, when the group was assessing the sound level it was split into two subgroups with slight differences of opinion as to which spectral balance they preferred. The results, however, were not conclusive since unintended variations of other parameters influenced the judgements.

Recently, the spectral balance has been investigated (by Lehmann) by applying octave values of EDT v. frequency (for the five octaves 125–2000 Hz). Attempts have been made to introduce the slope of these frequency curves (in s/octave) and to correlate this slope with judgements of musical samples from certain actually tested halls, e.g. by applying the semantic term 'hell' ('bright').

No doubt, the investigations of tonal quality and spectral balance are still in a very early stage and it is doubtful whether the question of differences in taste will ever be resolved or whether there will be established two, or more than two, groupings of different musical taste with regard to tonal quality.

Where does this research on criteria and correlation with subjective evaluation lead us?

Bearing in mind that three factors of assessment already account for 80–90% of the total variance it seems appropriate not to work with more than three criteria. We may choose, for example, C and RR plus something to express tonal quality. Another possibility which should be considered is to retain EDT (as a criterion for reverberance) and to supplement it with LE (as a criterion for early, lateral sound). When considering tonal quality the frequency dependence of EDT (octave values 125–4000 Hz) is a possibility in both cases.

Before making any choice we must also consider the advance testing of the various criteria in models as well as the final testing in completed auditoria. We have already encountered certain problems in applying the short time criteria, C and RR when testing in models and comparing with measurements in full scale halls, and we have considered increasing the frequency range to 2 or 3 octaves to avoid the random variations of the parameters. This means that our model simulation must also be more meticulous and must include a correspondingly larger frequency band.

In any event, the testing of tonal quality will only be relevant in finished halls (as long as we do not pass on to the elaborate methods of reproducing the complete frequency range in the models together with 2-channel recording and listening to musical samples). But what are the possibilities in general of testing in finished halls? The pulse method is easily applicable in empty halls but difficult to apply in occupied halls. The attempts to employ the orchestra as a sound source (with an audience present) have so far only been tried for the parameters RT and EDT.

Everything considered, it seems quite reasonable to continue working with EDT as a criterion for reverberance and with LE as a criterion for lateral reflections. The total balance may then (tentatively) be expressed by the EDT v. frequency dependence. This also means that we still consider valid the results of the early experiments in broadcasting studios (Chapter 1) in view of the fact that EDT and RT are identical in medium sized studios.

So far we have omitted any reference to a much discussed acoustical criterion, namely the so called *interaural coherence* (IC). In our opinion it is not necessary to consider this criterion as an independent parameter since it is closely associated with LE. Gottlob has shown that, under certain conditions we may express IC as:

$$IC \sim \text{early nonlateral energy/early total energy} \sim 1 - LE$$

Most likely the numerical values of LE will depend upon the method we apply when measuring the early, lateral energy. The method we have chosen here (to apply a figure-of-eight microphone) will appear to give numerical values different from those obtained by other methods. This seems a minor problem which may be overcome by establishing tables of conversion.

Some recent publications of W. Kuhl have put further emphasis on the vital importance of lateral reflection. Working with the subjective evaluation of steps of 'Räumlichkeit', in association with a number of concert halls, Kuhl has established a dependence of 'Räumlichkeit', correlated with the level of the orchestra sound and with the amount of lateral reflection. Kuhl uses the amount of lateral energy in the interval 10–80 ms in proportion to the direct energy (say the 0–5 ms interval).

According to their various shapes and energy levels the concert halls considered show considerable difference in 'Räumlichkeit' (rectangular halls with side balconies showing the highest levels). It thus seems justified to assume that the concept of lateral efficiency (LE) is well suited to use in the same context, notwithstanding that numerical values must depend upon which definition is used when talking of 'amount of lateral reflection'.

Acceptable Values

The various criteria discussed in this chapter are by no means firmly established when it comes to the question of acceptable values and tolerance ranges. All the same, some tentative values for concert halls are given in Table 11.1.

TABLE 11.1
ACCEPTABLE VALUES OF CRITERIA
APPLIED TO CONCERT HALL ASSESSMENT

Criterion	Average	Tolerance range
RT (s)	2·1	1·4–2·8
EDT (s)	2·1	1·8–2·6
C (dB)	0	−2– +2
RR (dB)	0	−0·5– +0·5

Lateral efficiency, LE, has not yet been extensively applied but it is a sensible guess to assume that values below 0·3–0·4 are unacceptable. The tolerance ranges of RT and EDT are taken from the results of Eysholdt. The average value for C is chosen to be zero, as recommended by Reichardt and Abdel Alim. Their recommended tolerance range, −1·6–+1·6, has been increased somewhat. This is motivated by the fact that there is a certain ambiguity in the choice of C = 0 as the normal value, since they found evidence that this value may depend upon the style of the composer. The tolerance range for RR is very tentative but it is in accordance with preliminary experiences from our own measurements.

Having thus surveyed the problems of criteria we shall continue the model testing using the criteria EDT and LE, but also occasionally taking the criterion C into consideration.

Concert Hall, Odense
We have already mentioned three cases where measurements of LE were performed (Chapter 10) and we have also compared the measured (average) values with the expected values (calculated from a formula given in Appendix V). We shall now look at a current project for a new concert hall in Odense of which a 1/10 scale model has been tested by measuring EDT, LE and C.

The concert hall is going to function largely as a concert hall, the home of the Symphony Orchestra of Fyn, with a secondary function as a congress hall. As in the case of the Oslo Concert Hall the requirements of congress

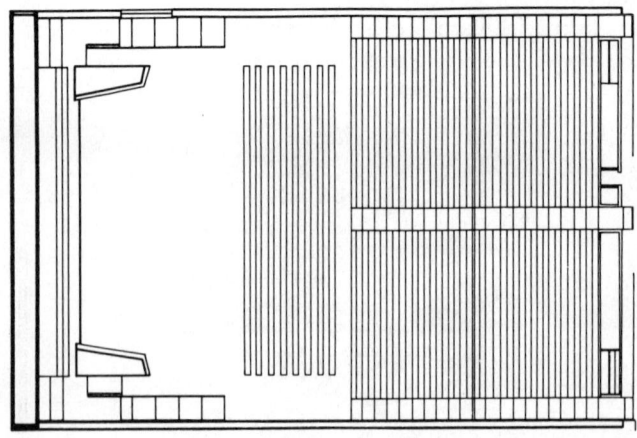

Fig. 11.1.   Concert Hall, Odense—plan.

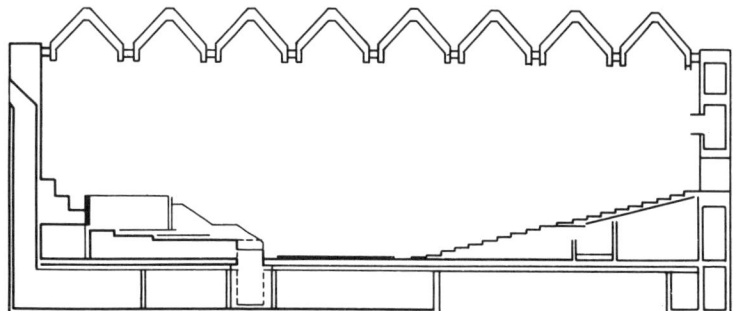

FIG. 11.2.   Concert Hall, Odense—longitudinal section.

have to be taken care of by the design of a high quality public address system so that the acoustical design of the concert hall for symphonic music will have overriding preference.

With a capacity of around 1200 seats it was decided to use a simple rectangular shape without any balconies, but with a terraced section at the rear. Modifications of shape may be introduced around the stage end where a seating area for choir (or audience) is included (Figs. 11.1, 11.2 and 11.3). As structural elements for the roof structure it was thought economic to use prefabricated elements of a triangular shape (in cross section). This shape

FIG. 11.3.   Concert Hall, Odense—view of the model.

was considered to be quite acceptable acoustically since it would provide plenty of diffusion and also in general character have some resemblance to the concept of 'beams across' (see Chapter 4). It was agreed that the gross shape of the interior ceiling may have to be supplemented with vertically oriented baffles to increase ceiling diffusion.

The main dimensions in plan are $29 \times 42\,m^2$ whereas the free ceiling height would be finally fixed according to the dictates of the model testing. A variation of the free ceiling height (maximum) between 16 and 14 m did not seem to be particularly critical for values of LE. For practical reasons the height may end up by being fixed by other considerations between these two limits. Looking at the dimensions in plan it may be argued that a somewhat smaller width may have been advantageous but the amount of lateral reflection may be more influenced by the distance between the platform (parapet) walls and the freely suspended side reflectors above the stage.

Preliminary measurements of LE seem to indicate that in some instances, average values of LE may exceed expectation values, a result which is in accordance with the experience gained in other models of rectangular halls. (For Dublin Concert Hall and Umeå Opera Theatre values see Table 10.1.)

Recent testing of this model has led to some interesting results. Values of C, generally, are somewhat higher than expectation values which may be associated with the large dimensions of the 'breaks' in the ceiling (Fig. 11.2). Values of LE are given in Table 11 of Appendix V.

Speaking of the large 'breaks' in the Odense Concert Hall ceiling design, it may be interesting to dwell for a moment on the many instances in which large diffusing elements have been referred to as a particularly profitable design feature in different chapters of this book. Large vertical baffles were mentioned in connection with an Utzon design of the Sydney Concert Hall as well as in the ceiling design of the Latin American Halls. 'Beams across' were applied in the early design of the Oslo Concert Hall. The remodelling of the University Hall in Dublin retained the feature of having a number of pillars as diffusing elements on all walls.

Diffusing elements may be considered as belonging to the boundaries of a hall but may also appear as semi-independent decorative units, for example as rings suspended above the platform as in the final design of the Sydney hall, and the question could be posed 'Is there perhaps no limit to the usefulness of diffusing elements?'

Recently J. Bosquet has referred to some examples of classical halls where artists were responsible for interior decoration, resulting in an abundance of diffusion. He also refers to the investigations of H. Kuttruff on the application of diffusing elements in reverberation rooms. If it may be regarded as an ideal for concert halls to obtain build-up and decay processes which very rapidly approach statistical conditions (as indeed it has been stressed in this book) it is hardly possible to exaggerate the amount and extension of diffusing elements, which could and should be applied. The suggestion (of Bosquet) to restore the role of the artist and to establish a cooperation between architect, artist and acoustical engineer may well be worth considering.

BIBLIOGRAPHY

Subject	*Reference*
Subjective judgement factors	Siebrasse, K. F., Doctoral Thesis, University of Göttingen, 1973.
Correlation between subjective judgement factors and objective criteria	Gottlob, D., Doctoral Thesis, University of Göttingen, 1973.
Reverberance and spatial effect	Eysholdt, U., Gottlob, D., Siebrasse, K. F. and Schroeder, M. R., DAGA '75, VDI Verlag, Berlin-West, p. 471.
Early lateral plus reverberant energy correlated with preference scale	Gottlob, D., Siebrasse, K. F. and Schroeder, M. R., DAGA '75, VDI Verlag, Berlin-West, p. 467.
Correlation between subjective and objective criteria	Lehmann, P., DAGA '72, VDI Verlag, Berlin-West, p. 162.
Semantic scale applied by subjective assessment of musical samples	Wilkens, H., Doctoral Thesis, Technical University of Berlin, 1975.
Subjective investigations and digital simulations of sound fields from concert halls	Eysholdt, U., Doctoral Thesis, University of Göttingen, 1976.
Spectral balance	*ibid.* Lehmann, P., quoted by Cremer, L., in *Die Wissenschaftliche Grundlagen der Raumakustik*, Vol. I, p. 484, S. Hirzel Verlag, Stuttgart, 1978.
Interaural Coherence v. Early non-lateral energy/Early total energy	Gottlob, D., Doctoral Thesis, p. 54, University of Göttingen, 1973.

Acceptable values of:
RT

Eysholdt, U., Doctoral Thesis, University of Göttingen, 1976.

EDT

ibid.

C

Reichardt, W., Abdel Alim, O. and Schmidt, W., *Acustica*, **32**, 1975, p. 126.

'Raumlichkeit' as related to the amount of lateral reflection

Kuhl, W., *Acustica*, **40**, 1978, p. 167.

Acceptable values and measured values of LE

3rd Symposium of the Federation of Acoustical Societies of Europe on Building Acoustics, 1979, Dubrovnik, Yugoslavia (Yugoslav Committee for ETAN, Beograd, Kneza Milosa 9).

Summary of research on acoustical criteria

New Aspects of Room Acoustics, Nederlands Akoestisch Genootschap Publication No. 49, 1979.

Gottlob, D., Neue Aspekte in der Konzertsaalakustik (pp. 44–62).

The usefulness of diffusing elements

ibid.

Bosquet, J., *Acoustique des Salles*, section II, pp. 74–8.

Kuttruff, H., *Acustica*, **18**, 1967, p. 132.

Chapter 12

Natural and Artificial Acoustics

To sit in a concert hall and enjoy the music is one possibility, enhanced if the natural acoustics of the hall is such that it cooperates with the conductor and the orchestra. To sit at home and enjoy a transmission, or a tape or gramophone record is another possibility and here the essence of the problem is whether or not the cooperation of the natural acoustics of the concert hall with the music can survive the artificial link connecting the hall to the listener in his home.

Fifty years ago everything was very simple: a primitive microphone picked up the music played in an equally primitive studio and a rather narrow frequency band was reproduced over an even more primitive loudspeaker. Even twenty-five years ago the radio channel only transmitted mono signals and although a few concert studios had been built and the frequency range of the whole chain from the orchestra to the listener had been improved tremendously the radio transmission was still something which could be compared with having one ear close to a small hole in one wall of the studio, as mono transmission has been repeatedly described.

The all important event in the development of transmission (including recording/reproducing) is the introduction of stereo links (two channel transmission), which, in principle, created the aural illusion of sitting in the same room as the musicians. Of course, there are certain limitations inherent in the substitution of a large concert hall or orchestra studio with a drawing room of normal dimensions. Low frequency resonance may influence the frequency response and so may the reverberation time of the room.

The nearly perfect transmission still requires that the listener use high quality headphones so that, in principle, the reproduction of the sound picture in the concert hall is feasible. The microphone arrangement in the concert hall must then be the system introduced under the name 'Artificial Head', with miniature microphones inserted in the two ear canals of a dummy head. It is essential that the dummy head has the correct dimensions and appearance of a normal head. With this arrangement and with each microphone connected (through a two channel transmission) to each of the two headphones it is possible to come close to an ideal simulation. The localisation of the orchestra is not perfect, particularly the groups close to a central position which produce a sound inside one's head instead of in front of it. Certain improvements may still be possible but will require automatic compensation for the random movements of the listeners' heads which is an added complication.

If we accept the drawing room with its imperfections as the receiving room and have a normal stereo setup with a proper distance between two loudspeakers and if we retain the artificial head in the orchestra studio we can still experience 'the presence' in the studio with normal reverberance of the performance.

The development of stereo reproduction into what has become known as quadrophony does not seem to be justified because the lateral orientation of the orchestra is sacrificed and abnormally high reverberance results. These two deficiencies are not compensated by the sense of 'being surrounded by the music' which seems to be the selling argument of quadrophony. To feel present in the concert hall is still preferable to the sensation of being inside the orchestra.

So far, we have only treated the microphone pick-up arrangement in the studio as if we were using an 'artificial head' system. Normally, the artificial head would be located at an ordinary audience seat at a considerable distance from the orchestra platform.

The currently applied methods of stereo pick-up are either coincidence microphones (combination of omnidirectional and figure-of-eight microphones) or a pair of directional microphones (20–50 cm apart), in both cases suspended close to the front of the orchestra. These two methods are both inferior to the artificial head arrangement when it is intended that reproduction should take place via a stereo loudspeaker arrangement in a (reasonably well damped) drawing room. The sense of reverberance is increased to more than normal but the sense of being present in the concert studio is less than that produced by the artificial head pick-up since the studio is located as though it were *behind* the loudspeakers.

This is an argument which is definitely in favour of developing the artificial head method for microphonic pick-up for studios as well as concert halls, for radio transmission and for recording.

Principally, an artificial head with built-in microphones produces the exact sound picture of what happens, musically and acoustically, at the particular position of the artificial head in the concert hall. This means that direct sound, early reflections and reverberant sound are transmitted (by two channel transmission) to the listeners' two stereo speaker systems. The fact that one listens with both ears to both speakers does alter the sound picture somewhat but this may be compensated by electrical filters. The position of the listener is however critical and he must sit exactly between the two stereo loudspeakers.

If, at the particular location of the artificial head, the concert hall has satisfactory values of the main criteria then a satisfactory acoustical performance would be expected in the drawing room with the important characteristic features of that particular concert hall (at that particular location) retained.

A concert hall which has satisfactory values for the main criteria may be analysed by such a set of criteria values. It may also be characterised by the corresponding oscillogramme of a short impulse giving the amplitudes of direct sound, early reflections and reverberation.

When working on extended research concerning clarity (transparency) and room impression ('Räumlichkeit') W. Reichardt *et al.* examined a large number of synthetically produced sound fields (with direct sound, early reflections and reverberation and with variation of amplitude, time delay and angle of incidence for all components). Out of 35 different samples it was possible to make subjective assessments and to arrive at one particular sound field as representing optimum conditions, i.e. optimum with regard to both clarity and room impression.

This optimum sound field, which is shown in Fig. 12.1, had the

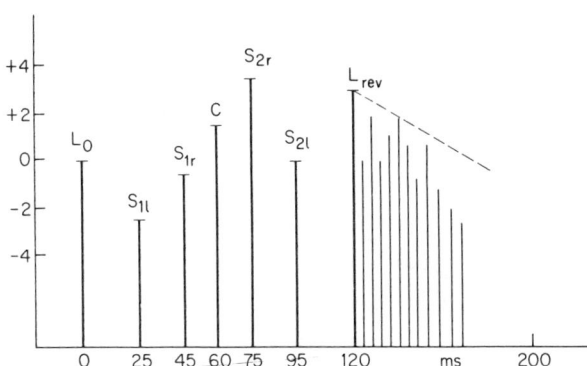

Fig. 12.1.   Schematic reflectogram of optimal character.

TABLE 12.1

EXAMPLE OF OPTIMUM SOUND FIELD CONDITIONS FOR CONCERT HALLS

Component	Delay (ms)	Level (dB)	Angle of incidence (degrees)
Direct sound	0	0	±25
1st left, lateral reflection	25	−2·5	35
1st right, lateral reflection	45	−0·5	50
Ceiling reflection	60	+1·5	50
2nd right, lateral reflection	75	+3·5	120
2nd left, lateral reflection	90	0	78
Reverberation	120	+3·0	±45, ±135

components given in Table 12.1. Looking at the figure we notice the close grouping of reflections within the interval 25 to 80 ms. As we know, it is within this interval that lateral reflections cause an increase of both parameters, C as well as RR.

If we transform the oscillogramme (of Fig. 12.1) to the integrated response we can also graphically illustrate the pulse picture (either linearly, Fig. 12.2 or in terms of decibels, Fig. 12.3). From these illustrations we can deduce some of the other criteria, with which we have been concerned, namely steepness (σ) and rise time (TR). The approximate values are:

$\sigma \sim 0\cdot1$ dB/ms,

TR $\sim$ 70 ms

both of which may be classed as outstanding values, especially the value of steepness. Furthermore, it can be inferred from the pulse pictures that

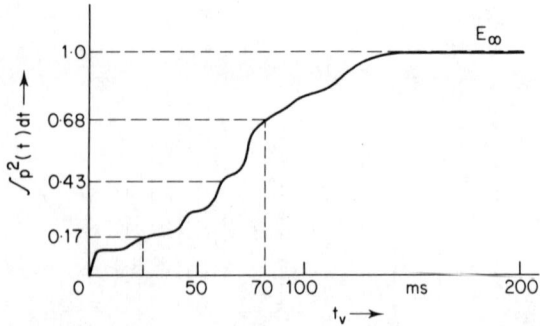

FIG. 12.2.   Integrated response of reflectogram, linear scale.

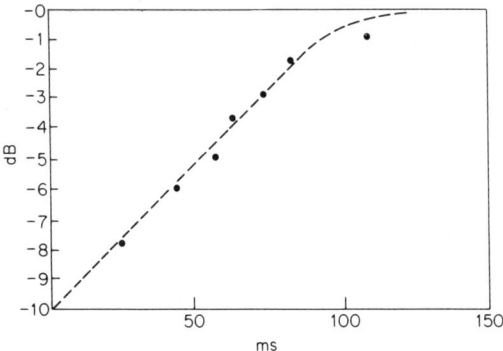

FIG. 12.3.   Integrated response of reflectogram, logarithmic scale.

lateral efficiency will have a very high value (LE ∼ 0·7–0·8) and that clarity also will be high (around +2 dB).

The question arises whether it is possible to obtain, in actual halls, sound fields approaching such ideal conditions. What looks extraordinary is, of course, the levels of some of the lateral reflections which are higher than the level of the direct sound.

As an example of a modern concert hall where the acoustics is exclusively artificial it is worthwhile mentioning the well known 6000 seat Congress Hall of Moscow. In spite of the enormous size of the hall the natural RT has been reduced to around 0·9 s and a very elaborate system of loudspeakers located at the side walls and in the ceiling (behind perforated surfaces) radiates the direct sound plus a number of delayed reflections and reverberated sound. The microphonic pick-up is by a long row of directional microphones located along the proscenium edge at stage level. Although this system dates back to the early sixties the performance is of very high quality and the arrangement is sufficiently sophisticated to give a close to perfect illusion of natural acoustics at most locations of the audience seating. The exception is the first rows close to the stage where the blending of the natural sound and the reinforced sound is not complete. The hall is used for opera as well as for concerts.

The new knowledge gained by the experiments of Reichardt *et al.* concerning the composition of optimum sound fields may well serve the purpose of developing both the art of artificial acoustics and the art of natural acoustics and here we shall briefly consider both possibilities, especially in connection with improvement schemes. When working with artificial acoustics it will always be preferable to reduce the natural reverberation to a minimum (as in the Moscow hall) and to arrange the electro-acoustic system to completely dominate the acoustics of the hall. This is not feasible in existing halls and a problem which will present itself quite often is the following: given an existing hall with some acoustic deficiencies, is it possible by adding an electro-acoustic system (a system which has a very high operational safety) to eliminate these deficiencies to a reasonable extent? Or, is it preferable to go into the more elaborate (and less discrete) scheme of changing surfaces, arranging reflectors etc. to achieve acoustical improvements?

There will be no general answer to these questions which will have to be answered in a way specific to each case. We have already mentioned the case of the Malmö City Theatre where improvement will include a change of shape to the orchestra pit, an extension of the ceiling height and also changes of several surfaces. We may, as in another case, consider the possibilities of improving the acoustical properties of the Concert Studio of Radiohuset, either by adding an artificial system or by changing some of the surfaces.

The scope of an artificial system would be to increase the amount of lateral energy. It is not possible to do this over all three seating areas. The seating on the orchestra level is too close to the stage so that any arrangement of speakers and microphones will have a dangerously low limit of feedback level. The seating on the second balcony does not need much improvement because of the proximity to the rear wall. The seating on the first balcony would be the appropriate area to choose for an improvement of lateral energy. Microphonic pick-up for the system would be provided from a pair of hyper-cardioid microphones (20 cm apart) centrally located in front of the platform. The signals would have two delays adjusted so that the total delay for a central seat position on the first balcony would be, for example, one of 35 and 55 ms, on the one channel (left) and one of 40 and 60 ms, on the other channel (right).

The speakers should be horizontally orientated sound columns (placed as shown in Fig. 12.4). With a length of, for example, 4 m, the columns would have a small opening angle in the vertical plane through the centre of the column axis, thus generating a minimum of feedback to the microphones. Only an experimental setup of such a system can prove whether or not there is an improvement.

The possibility of changing certain shapes in the concert studio would create the opportunity of increasing the amount of lateral energy. This may be achieved in several ways. One is to work with extended systems of reflectors protruding from the side walls and angled to direct the reflected energy towards the seating areas. Most likely they would have to be angled in both main directions which would make them look rather conspicuous in the hall. Even so, their effect in the centre area of the seating might be rather limited due to the large distances.

Another possibility is to work with suspended vertical baffles in much the same way as was done in the National Theatres of Latin America (Chapter 5). One of the (assumed) advantages of vertically oriented baffles is that they

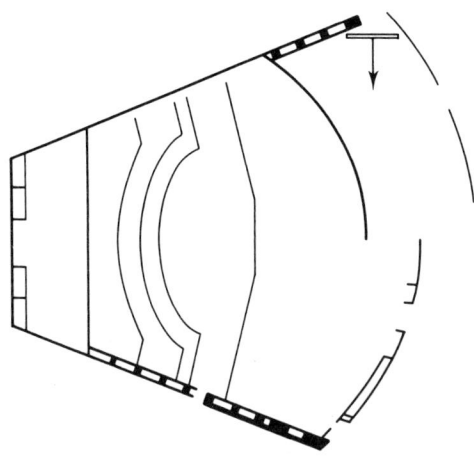

FIG. 12.4.   Suggested artificial system for Concert Studio.

tend to feed more energy into horizontal directions during a process of many repeated reflections. Of course, in both cases of improving the acoustics, the artificial as well as the natural, careful testing of the main criteria (especially of LE) is vital, both before and after the introduction of any changes. Only by such a procedure will it be possible to get any objective improvements documented.

It is appropriate that these considerations concerning the new vistas (opened up by the results of acoustical criteria research over the last 20 years) should end on an optimistic note. The mere fact that we have a considerable knowledge about the ideal range of the main criteria for a concert hall (or a concert studio for that matter) is an excellent starting point for future design work, not only when considering new projects but also when working on the improvement of existing halls.

It is also good to know that some of the main criteria have considerable tolerances in numerical values and that others may depend partly on subjective preferences or on musical characteristics. We have already mentioned the case of clarity where the optimum value may depend upon the àctual musical piece performed. Recently it has been found by experiment that, for example, the delay of the first (single) reflection should ideally vary within wide limits according to the nature of the musical score. On the other hand, the experiments also showed that the angle of incidence for this first (single) reflection may vary between limits of $55° \pm 20°$ and yet result in optimum hearing conditions (irrespective of which one of the four musical scores was used). Therefore the new numerical criteria as well as graphical representations of the experimentally verified optimum reflection patterns, may be thought of not as rigid standards to be strictly complied with, but rather as valuable guidelines which in many cases may ease the work of the acoustical designer and make him rely less on intuition and more on exact knowledge. Even so, there will still be a wide range of design work for him on which to use his imagination in collaboration with architects and other consultants.

Looking into the foreseeable future, this design work will still have a need for the use of scaled models. Therefore, the long development of model research reported on these pages will continue. No doubt, the testing of physical models will be supplemented by, for example, computer models in which there is growing interest. Such models will be useful for calculating approximate values of the main criteria on the basis of the gross shape of a hall. They will be insufficient as soon as we are working with all the finer details of shape. The sound diffusion from surfaces and from obstacles is an extremely complex process, too complex even for a computer, but not for a scaled physical model.

BIBLIOGRAPHY

Subject	*Reference*
Stereophony, quadrophony, artificial head, etc.	Kuhl, W. and Plantz, R., *Rundfunktechnische Mitteilungen*, **19**, 1975, pp. 120–132.
Listening over headphones to different kinds of microphone pick-up	*Ibid.*
Compensation for head movements when listening over headphones	Boerger, G. and Kaps, U., DAGA '73, VDI Verlag, Berlin-West, pp. 398–401.

Clarity and room impression, subjective assessment of synthetically produced sound fields

Reichardt, W., Abdel Alim, O. and Schmidt, W., Z. Elektr. Inform.-u. Energietechnik, **5**, 1975, (2), pp. 144–155.

Delay and angle of first (single) reflection

Ando, Y., DAGA '76, VDI Verlag, Berlin-West, pp. 259–262.

Head related two-channel stereophony with loudspeaker reproduction

Damaske, P., J.A.S.A., **50**, 1970, pp. 1109–1115.

Appendix I

BUILDING-UP PROCESS, RISE TIME AND INVERSION INDEX

It is assumed that the decay process in a hall follows an exponential function:

$$I_t = I_0 e^{-kt} \tag{1}$$

(I_t intensity, I_0 intensity at-zero time)

$$k = cA/4V \tag{2}$$

(c, velocity of sound, A total absorption, V volume)

$$RT = 0.16V/A \tag{3}$$

then from (2) and (3):

$$k = 13.76/RT \tag{4}$$

A corresponding (complementary) build-up process may be written:

$$I_t = I_0(1 - e^{-\frac{13.76t}{RT}}) \tag{5}$$

The value of rise time (TR) will correspond to the point of time where 50% of the total energy has arrived:

$$I_{TR}/I_0 = 0.5$$

hence:

$$e^{-\frac{13\cdot76\text{TR}}{\text{RT}}} = 1 - I_{\text{TR}}/I_0 = 0\cdot5$$

$$-\frac{13\cdot76}{\text{RT}}\cdot\text{TR}\cdot\log e = -0\cdot3$$

$$\text{TR} \cong 0\cdot05\text{RT (s)} \tag{6}$$

The inversion index (II) may be defined as:

$$\text{II} = \frac{\text{TR (average in audience area)}}{\text{TR (average in stage area)}}$$

Appendix II

BUILDING-UP PROCESS, STEEPNESS AND INVERSION INDEX

Assuming (as in Appendix I) a build-up process which is complementary to the decay process:

$$I_t = I_0(1 - e^{-\frac{13 \cdot 76t}{RT}}) \tag{7}$$

hence:

$$10 \log \frac{I_t}{I_0} \sim 10 \log (1 - e^{-\frac{13 \cdot 76t}{RT}}) \tag{8}$$

The slope of this curve may be written:

$$\frac{d}{dt}\left(10 \log \frac{I_t}{I_0}\right) = \frac{d}{dt}[10 \log (1 - e^{-\frac{13 \cdot 76t}{RT}})] \quad \text{(dB/s)} \tag{9}$$

or, expressed in dB/ms:

$$\frac{d}{dt}\left(10 \log \frac{I_t}{I_0}\right) = 10^{-2} \log e \frac{\frac{13 \cdot 76}{RT} e^{-\frac{13 \cdot 76t}{RT}}}{1 - e^{-\frac{13 \cdot 76t}{RT}}} \quad \text{(dB/ms)} \tag{10}$$

At a level $-5 \, \text{dB}$ below the stationary level, we get:

$$\sigma_{\text{calc.}} = \frac{d}{dt}\left(10 \log \frac{I_{t(-5\text{dB})}}{I_0}\right) = 0 \cdot 0094 \frac{13 \cdot 76}{RT} \simeq \frac{0 \cdot 13}{RT\,(\text{s})} \quad \text{(dB/ms)} \tag{11}$$

208

TABLE 1

MEASURED VALUES OF TR AND σ IN CONCERT STUDIO, RADIOHUSET, EXPRESSED AS A PERCENTAGE OF CALCULATED VALUES

$TR_{calc.} = 100\,ms \qquad \sigma_{calc.} = 0.065\,dB/ms$ (empty)

$\sigma_{calc.} = 0.087\,dB/ms$ (audience simulated)

Location		Empty hall, with reflectors TR (ms)	Empty hall, without reflectors TR (ms) σ (%)		Audience, without reflectors σ (%)
Sound source	Microphone				
Stage right	Stage left	48	70	94	25.0
Stage front	Stage rear	55	105	30.5	24.0
Stage right	Floor, 6th row	70	60	95.5	51.5
Stage right	1st balcony, 1st row	60	50	115	83.0
Inversion Index		1.26	0.63	0.57	0.36

Comment: note the great difference of II with and without reflectors. Also note the detrimental influence of the audience absorption on the II.

Further, we can define the Inversion Index (II) as:

$$II = \frac{\sigma \text{ (average value at the stage)}}{\sigma \text{ (average value in the audience area)}} \qquad (12)$$

Values of rise time, steepness and II for the Concert Studio, Radiohuset, are shown in Table 1.

TABLE 2

MEASURED VALUES OF STEEPNESS IN NEW YORK STATE THEATRE (%)

Microphone position†	Model		Theatre	
	$RT = 0.19\,s$	$RT = 0.16\,s$	$RT = 1.9\,s$	$RT = 1.6\,s$
Orchestra level, centre	63.6	65.8	60.9	76.4
Orchestra level, rear	76.9	101	81.2	105
Orchestra level, side	112	123	81.2	145
Orchestra level, between centre and side	97.5	89.8	81.2	107
4th balcony, centre	97.5	82.1	69.6	82.1
Average	89.5	92.3	74.8	103
$\sigma_{calc.}$ (dB/ms)	0.684	0.813	0.0684	0.0813

† See Fig. 4.4.

TABLE 3
PERCENTAGE VALUES OF STEEPNESS IN MODEL OF 'MAJOR HALL'

Microphone position	Model conditions		
	Audience simulation:		With baffles
	Absorbing mats	Neoprene	
Stage, left	65	94	120
Stage, centre	82	97	112
Stalls, front	81	89	106
Stalls, centre	46	59	80
Stalls, side	60	61	79
Terrace, centre	75	87	105
Terrace side, rear	77	90	90
Terrace side, front	64	93	97
Average	68·8	83·8	98·6

Another expression for steepness may be written:

$$\sigma \cong 0 \cdot 0094 \frac{c\alpha_{\mathrm{m}}}{p} \tag{13}$$

where c is the velocity of sound, α_{m}, average absorption coefficient and p, mean free path. Equation (13) follows from an expression for the decay constant k:

$$k = \frac{c\alpha_{\mathrm{m}}}{p}$$

Measured values of steepness in a $1/10$ model of the New York State Theatre and in the completed theatre are indicated in Table 2. Values are stated as a percentage of calculated steepness.

TABLE 4
PERCENTAGE VALUES OF STEEPNESS IN $1/10$ MODEL OF
METROPOLITAN OPERA AND IN THE COMPLETED AUDITORIUM

Microphone position	Model	Auditorium
3–5 stall locations (average)	88	96
2–3 balcony locations (average)	101	104
Inversion Index	0·87	0·93

Table 3 gives measured values of steepness in 1/10 model of 'major hall' (Sydney Opera House project), and Table 4 gives measured values of steepness in 1/10 model of Metropolitan Opera, New York and in the completed opera auditorium.

Appendix III

DECAY PROCESS, EDT AND INVERSION INDEX

If the decay process follows the normal exponential function (Appendix I), values of EDT are theoretically identical to values of RT.

Measured values of EDT may, however, show deviations in both directions. According to our evaluation practice (established over the years) it has been usual to consider values *above* RT values and values *equal to* RT values as satisfactory, but values *below* RT values (especially if more than 15–20 % lower) as unsatisfactory.

This practice may have to be revised since recent research (discussed in Chapter 11) indicates that a valid criterion would be numerical values of EDT which are within a certain definite interval.

If methods existed which allowed calculations of values of the mean free path over a limited time range (corresponding to sound paths between source and receiver) then these values could be used to calculate EDT (and steepness).

When applying values of EDT to the calculation of inversion index the definition is as follows:

$$II = \frac{EDT \text{ (average value in audience area)}}{EDT \text{ (average value in stage area)}}$$

The following tables give numerical examples of measured values in models and halls of EDT and calculated values of II (Tables 5–8). An experiment of measuring directional values of EDT is recorded in Table 9.

TABLE 5

EXAMPLES OF MEASURED VALUES OF EDT(s) IN 1/10
MODEL AND IN AUDITORIUM OF NEW METROPOLITAN
OPERA, NEW YORK

	Microphone location	
	Model	Auditorium
EDT (average of 5 locations)	0·153	1·83
RT (average of 5 locations)	0·141	1·87

TABLE 6

CALCULATED VALUES OF II FROM
MEASUREMENTS IN 1/10 MODEL OF OSLO
CONCERT HALL WITH DIFFERENT
CEILING SHAPES

Ceiling shape	II calculated from	
	Steepness	EDT
Flat	0·78	0·90
Transverse beams	1·10	1·10
Radial beams	0·87	0·92

TABLE 7A

INVERSION INDEX FOR THE OPERA THEATRE, SYDNEY OPERA HOUSE, MODEL 1/10

Design	Sound source location			
	Pit	Stage	Pit	Stage
	(16 kHz octave)			
Flat ceiling	1·15	0·90	—	—
Stepped ceiling	1·13	0·97	—	—
High ceiling, balconies	1·26	1·15	1·17[a]	1·11[a]
Final design	1·27	1·10	1·22[b]	1·13[b]

[a] Average of octaves 8 and 16.
[b] Average of octaves 2, 4, 8 and 16.

TABLE 7B
INVERSION INDEX FOR THE OPERA THEATRE, SYDNEY OPERA HOUSE

Conditions	Sound source location			
	Pit	Stage	Pit	Stage
	(2 kHz octave)		(Average of octaves 125–4 000)	
Empty theatre	1·52	1·09	—	—
Capacity audience	1·28	1·15	1·22	1·10

TABLE 8A
INVERSION INDEX FOR THE CONCERT HALL, SYDNEY OPERA HOUSE, MODEL 1/10

Design	16 kHz octave	Average of 2, 8, and 16 kHz
1st design		
with reflectors at box soffit level	1·10	—
without reflectors	1·01	—
2nd design		
with reflectors at box soffit level	1·19	1·11
with reflectors just below crown	1·13	1·05

TABLE 8B
INVERSION INDEX FOR THE CONCERT HALL, SYDNEY OPERA HOUSE

Conditions	2 kHz octave	Average	
		125–4 000	500–4 000
Hall, empty			
reflectors—at box soffit level	1·14	1·14	1·18
just below crown	1·10	1·09	1·11
Hall, capacity audience			
reflectors—at box soffit level	1·00	0·97	0·99
just below crown	1·04	1·05	1·05
10·5 m above stage	1·10	—	1·11

TABLE 9
OSLO CONCERT HALL, 1/10 MODEL.
EXAMPLES OF MEASURED VALUES OF EDT,
AVERAGE AND IN DIFFERENT DIRECTIONS
(16 kHz OCTAVE, RT = 0·147 s)

Location	Direction	EDT (s)
Stalls, centre	Omni	0·187
	Vertical	0·158
	Longitudinal	0·162
	Lateral	0·187
Stalls, side	Omni	0·174
	Vertical	0·168
	Longitudinal	0·167
	Lateral	0·186
Balcony, centre	Omni	0·180
	Vertical	0·184
	Longitudinal	0·178
	Lateral	0·181

Appendix IV

IMPULSE RESPONSE LEVEL AS ACOUSTICAL CRITERION

Occasionally, the *impulse response level*, measured at a number of locations in a hall, has been applied as a criterion of the sound distribution.

This was so in the case of the Ruben Dario National Theatre of Nicaragua in Managua. The results are shown in Table 10. It will be seen from this table that the variation of the impulse response level (in dB) (as well as of the values of EDT) is remarkably small. This may partly be ascribed to the system of vertical baffles suspended under the ceiling in the auditorium (as described in the text, Chapter 5).

TABLE 10

Source location	Microphone location	Impulse response level (dB)	EDT (s)
Stage, extreme right, front	Same, left	104	1·79
Stage, centre, front	Centre, rear	103·2	2·04
Stage, extreme right, middle	Left, middle	105·7	2·02
Stage, extreme right, middle	Left, rear	104·5	2·01
Stage, extreme right, middle	Centre, front	112·7	2·11
Average on stage			2·00
Centre, front	Row F, seat 28	103·8	2·14
Centre, front	Row M, seat 26	103·8	1·90
Centre, front	Row T, seat 25	101·5	1·88
Centre, front	Row J, seat 5	103·2	2·06
Centre, front	1st balcony	105·2	2·02
Centre, front	2nd balcony	103·4	2·00
Centre, front	3rd balcony	103·4	2·04
Centre, front	1st balcony, left	103·7	2·11
Average in auditorium			2·02

Appendix V

EXPECTATION VALUES OF C, RR AND LE, AND MEASURED VALUES OF LE

According to the definitions we have the following expressions for clarity (C), room response (RR) and lateral efficiency (LE):

$$C = 10 \log \frac{E^o_{0-80}}{E^o_{80-\infty}} \quad (dB)$$

$$RR = 10 \log \frac{E^\infty_{25-80} + E^o_{80-160}}{E^o_{0-80}} \quad (dB)$$

$$LE = \frac{E^\infty_{25-80}}{E_{0-80}} \quad (coefficient)$$

The expectation values, i.e. the values calculated when assuming complete statistical conditions during the decay process (and exponential decay function) may be expressed by the following formulae:

$$C = 10 \log \frac{1 - p_2}{p_2} \quad (dB)$$

$$RR = 10 \log \left(\frac{2}{\pi} \cdot \frac{p_1 - p_2}{1 - p_2} + \frac{p_2 - p_3}{1 - p_2} \right) \quad (dB)$$

$$LE = \frac{2}{\pi} \frac{p_1 - p_2}{1 - p_2} \quad (coefficient)$$

where:

$$p_1 = e^{-13 \cdot 8 \, . \, 0 \cdot 025/RT}, \qquad p_2 = e^{-13 \cdot 8 \, . \, 0 \cdot 08} \quad \text{and} \quad p_3 = e^{-13 \cdot 8 \, . \, 0 \cdot 16/RT}$$

Numerical calculations of LE, RR and C for values of RT between 1·25 and 2·8 s are shown in Fig. A.1. Values of LE have been measured recently in a number of halls and models. The measured average values as well as

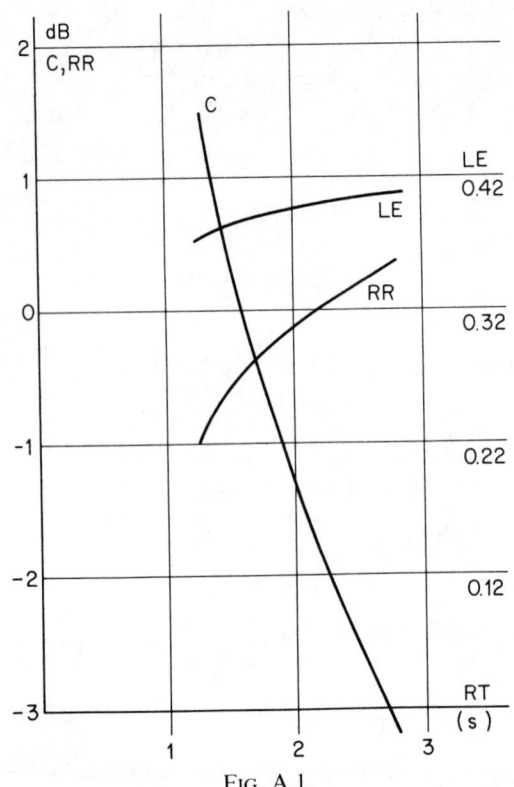

FIG. A.1.

minimum and maximum values have been calculated as percentages of the expected value (according to the formula above). The percentage values are given in Table 11.

The items refer to models and halls mentioned in Chapters 5, 7, 8 and 10 of this volume. The series I–V measured in the model of the Odense Concert Hall refer to the following conditions:

Series

I   No diffusing surfaces in the model,

II   Diffusers on the side walls,

III As (II) plus vertical baffles in the ceiling; distance between baffles 40 cm,

IV   As (III) plus freely suspended side reflectors to both sides of the stage,

V   As (II) but distance between baffles 20 cm. No side reflectors.

TABLE 11
MINIMUM, MAXIMUM AND AVERAGE VALUES OF LE MEASURED IN
VARIOUS MODELS AND HALLS, EXPRESSED AS PERCENTAGES OF
EXPECTED VALUES

Item	LE (%)		
	Average	Minimum	Maximum
Oslo Concert Hall			
without stage draperies	89·5	56·5	143
with stage draperies	93·0	21·9	124
Dublin Concert Hall			
model	127	93·5	280
Malmø Theatre			
model	64·5	35·3	108
opera	65·0	32·7	129
Odense Concert Hall			
model			
series I	85	5·05	175
II	96	14·8	172
III	108	14·9	195
IV	98	26·3	178
V	92	39·1	160
Linköping Concert Hall			
model	114	58	183
National Theatre of			
Guatemala	25·1	8·0	46·6

It is interesting to note the considerable influence of diffusing elements on the variation between minimum and maximum values of LE. A new project for the city of Linköping, Sweden has been tested in a 1/10 model. The design is for a rectangular hall with side balconies.

Index

Aalborghallen, Aalborg, 41–6, 54
Abidjan Congress/Concert Hall, 140
Absorption, 61, 62
Acceptable values, 191
Acoustical criteria, 57
Acoustical deficiencies, 9
Acoustical problems, 75
Acoustical properties, 6–8
Acoustical testing, 107
Altes Gewandhaus, Leipzig, 33, 34
Arena theatre, 139
Articulation, 134
Artificial acoustics, 39
Artificial head, 198, 199
Artificial support, 67
Arts cinema, 117
Assembly halls, 41, 45, 49, 54
Avery Fisher Hall, 83

Baffles, 84, 96
Ballet theatres, 63, 75, 83, 86
'Beams across' concept, 194
Berlin Theatre House, 149
Boston Symphony Hall, 35–7
Building-up process, 206, 208

Chamber music, 117
Cinema, 46, 49

Clarity (C), 131, 159, 160, 169, 172,
    175, 176, 179–82, 187, 190, 192,
    217
Concert
    halls, 12–18, 41, 45, 46, 49, 54, 55,
        57, 58, 83, 86, 92, 98, 100, 106,
        110, 113, 118, 119, 151, 175
    studios, 5–12, 59, 60
Concertgebouw, Amsterdam, 35
Congress Hall, Moscow, 201
Criteria, 158–64, 187–96

Decay
    criteria, 158
    process, 206, 208, 212
'Deutlichkeit', 58, 158, 159
Diffusion, 194, 206
Drama theatres, 77, 78, 98, 115, 121,
    151, 175
Drying-out procedure, 61, 62
Dublin Concert Hall, 170–5, 183, 185,
    195, 219

Early Decay Time (EDT), 52, 55, 61,
    70, 71, 73, 80, 85, 88, 99, 100,
    103, 107, 109, 123, 124, 127, 129,
    130, 137, 146, 147, 155, 158, 168,
    172, 175, 179, 188–92, 212, 216

221

Early lateral energy, 159, 179
Echogrammes, 129
Electrical tapering, 113
Energy
    concept, 57
    criteria, 70, 158
Epidauros, 22

Frequency range, 61, 62
Frontal energy, 161

Glazing, 105
Greek amphitheatre, 22
Guatemala National Theatre, 85–8,
    91

'Hallabstand', 158
'Halligkeit', 188
'Hallmass', 158, 160
Headphones, 197
Helmholtz resonators, 8
Herodes Atticus, 22, 25
Historical survey, 22–38

Impulse response level, 216
Index of room impression, 160, 161,
    188
Initial reverberation time, 70
Instrumentation, 62
Interaural coherence (IC), 190, 191
Inversion, 11, 60
    Index (II), 11, 60, 61, 100, 103, 104,
    107, 123, 124, 127, 129, 146, 147,
    168, 172, 206, 207, 209, 212

Juillard School of Music, 83

Lateral efficiency (LE), 179, 181–4,
    187, 191, 192, 194, 217, 219
Lateral energy, 161, 202

Latin America, 83–8, 203
Level of sound, 188
Lincoln Centre, 83
Linköping Concert Hall, 219
Loudspeaker arrangements, 64, 77,
    88, 113, 202

Major Hall, Sydney, 63, 67, 92
Malmø City Theatre, 140, 149–51,
    178–85, 195, 202, 219
Metropolitan Opera House, New
    York, 30, 70–3, 78–83, 89
Microphone arrangements, 65, 104,
    130, 161, 169, 175, 179, 182, 198,
    202
Minor hall, Sydney, 96, 98
Model research, 62–73, 76, 79, 98,
    100, 103, 107, 122, 123, 129, 130,
    146, 158–64, 168, 172, 175, 179,
    184, 187–96, 219
Moscow, Congress Hall, 201
Multiform/multipurpose halls, 39–56,
    139–57, 175, 178
Municipal Theatre, Umeå, 175–8
Music rooms, 98, 117, 121
Musical criterion, 158, 159
Musikvereinsaal, Vienna, 35

National Theatre of Guatemala, 85–8,
    91
National Theatre of Nicaragua, 83–5,
    90, 216
National theatres of Latin America,
    203
Natural acoustics, 39, 197
Neues Gewandhaus, Leipzig, 34
New York State Theatre, 60, 63, 68,
    75–8, 83, 89

Odense Concert Hall, 192–5, 218, 219
Open stage concept, 140
Opera, 77, 81, 83, 86, 104, 118, 151
    theatre, 92, 100–5, 175, 176
Opera/drama theatres, 41, 44, 46, 54
Orange, France, 25

Organ, 6
Oslo Concert Hall, 71–3, 122–38, 192, 219
Oslo Recital Hall, 138

Palais de Congres, Houphouët-Boigny, 144–9, 156
'Point of gravity' time, 159, 160
Polar distribution pattern, 113

Quadrophony, 200

Radiohuset, Copenhagen, 1–12, 19, 44, 59, 202, 209
Räumlichkeit, 188, 191, 199
Reception hall, 117
Reflection, 159, 199, 203, 204
Reflectors, 108, 109, 177
Rehearsal/recording studio, 114, 115, 120
Renaissance theatre, 25
Resonance, 2, 57, 197
Reverberance, 188, 190
Reverberant/early energy, 158
Reverberation, 107, 197, 199
Reverberation time (RT), 1, 2, 3, 5, 8, 10, 11, 15, 16, 19, 20, 26, 33, 36, 44, 45, 46, 48, 52, 54, 55, 57, 58, 61, 62, 66, 71, 76, 77, 79, 80, 81, 85, 88, 99, 100, 103, 107, 109, 117, 129, 131, 132, 137, 141, 143, 146, 147, 168, 179, 192
Reykjavik University Hall, 49–53, 55–6
Rise time (TR), 11, 44, 59, 200, 206
Roman amphitheatre, 22–6
Room impression index (R), 187
Room response, 131, 161, 169, 172, 175, 176, 179–82, 188–90, 219
Royal Albert Hall, 57
Ruben Dario Theatre. See National Theatre of Nicaragua

Sabine reverberation time, 36
St Andrews, Glasgow, 35

Salle Pleyel, 6, 37, 58
Scala Opera, 30
Scala Theatre, Aarhus, 46–9, 55
Screening, 168
Spatial effect, 160, 161, 187, 188
Spectral balance, 188, 189
Spectral density, 189
Speech
    articulation, 59, 158
    criterion, 58
    quality, 134
    system, 135
Stadt Casino, Basle, 35
Steepness, 59, 61, 62, 66–73, 76, 80, 96, 200, 209–11
Stereo
    links, 197
    reproduction, 198
Stockholm Concert Studio, 165–70, 184, 195
Sydney Opera House, 63, 67, 92–121

Tape recorder, 62
Teatro alla Scala, Milan, 30
Teatro Colon, Buenos Aires, 30, 32
Teatro Farnese, Parma, 26
Teatro Olympico, Vicenza, 25, 26
Teatro San Carlo, Naples, 28–30
Test performances, 107
Thrust stage, 77
Tivoli Concert Hall, Copenhagen, 12–18, 20, 123
Transparency, 160, 187

Umeå Municipality multiform/multipurpose theatre, 140, 151–5, 175–8, 183

Vivian Beaumont Theatre, 83

Wagner Festspielhaus, Bayreuth, 32–3
Wintergarden Theatre, London, 140, 141–4, 156